The Theory of Fourier Series
and Integrals

The Theory of
Fourier Series
and Integrals

P. L. Walker

Department of Mathematical Sciences
University of Petroleum and Minerals,
Dhahran
Saudi Arabia

A Wiley–Interscience Publication

JOHN WILEY & SONS

Chichester · New York · Brisbane · Toronto · Singapore

Library of Congress Cataloging-in-Publication Data:

Walker, P. L. (Peter, L.), 1942–
    The theory of Fourier series and integrals.

    'A Wiley–Interscience publication.'
    Includes index.
    1. Fourier series.   I. Title.
QA404.W35   1986   515'.2433   85-17931

ISBN 0 471 90112 1

British Library Cataloguing in Publication Data:

Walker, P. L.
    The theory of Fourier series and integrals.
    1. Fourier series
    I. Title
    515'.2433   QA404

ISBN 0 471 90112 1

Photosetting by Thomson Press (India) Limited, New Delhi
Printed and bound in Great Britain

Contents

v

vi

Preface

This book has been written with two related aims, and with two groups of potential readers in mind. For students of mathematics, the first course in analysis is often an uncomfortable experience since they have to deal simultaneously with technical and conceptual ideas which are both new and awkward for a beginner to handle. It is difficult for a teacher to explain convincingly the rationale of the subject, and the students' dissatisfaction is further increased when 'Analysis I' is succeeded by 'Analysis II' (perhaps analysis in $\mathbb{R}^n$ or $\mathbb{C}$) without further ado. For these students, this book is intended to show how, with a rather small background in real analysis (which is outlined briefly below and in detail in Appendix A), one can obtain attractive and non-trivial results concerning Fourier series and integrals. On the other hand, for those who use Fourier theory in science or engineering, the theory is sometimes regarded as either irrelevant or too difficult to be studied. For them it is hoped the present account may show that one can go a great deal further, with very little in the way of prerequisites, than is generally believed.

Both these groups, it is hoped, will benefit from the introduction and illustration of the theory by means of its applications, particularly in the historically vital area of the conduction of heat: this interpretation not only motivates otherwise apparently arbitrary definitions or restrictions on the theory, but is an excellent source of illustration of the results. Both groups may also be stimulated to study the advanced theory for themselves; there is a combined bibliography and guide to further reading in Appendix C.

To be specific about prerequisites, we assume only that the student is acquainted with the notion of continuity for real or complex-valued functions of a real variable (and of several real variables in Chapter 6) and the notion of uniform convergence for sequences of such functions: we also assume sufficient knowledge of integration to enable bounded functions with only finitely many discontinuities (referred to throughout as FC-functions) to be integrated: in particular, the conventional Riemann theory is more than adequate here. All this information is summarized in Appendix A.

It should be abundantly clear from what has been said so far that there are also two aims which this book does *not* attempt to achieve. Firstly, it is not intended as an up-to-date account of the most technically complete results of the modern theory of Fourier series. There are many excellent books on the advanced theory, only a small number of which could be mentioned in the Bibliography, and there is a still wider collection of results in the mathematical literature: the present book is addressed specifically to beginners, though everyone will hopefully find at least an example which is new to them (a list of the author's favourites is given below).

Even more emphatically, this is not intended as a systematic account (or even an introduction) to the applications of the theory: such an undertaking would require a whole library rather than a single volume, and it is no part of our intention to attempt it. Our applications are included for the purpose of motivation and illustration, and the hope that students of pure mathematics might be encouraged to learn more about the applications, and vice versa.

It is this author's feeling that Fourier theory is singled out in particular by the wealth of interesting and elegant examples, which even more than the theorems themselves give the subject its individuality. A list of favourite theorems, for example 1.10, 1.19(i), 2.8(i), 2.10, 4.6, 5.18, 5.30, A2, and A30, is shorter than a list of favourite examples, such as 1.12(ii), 2.4(ii), 2.16, 3.22(iii), 4.3(i), 5.9(iii), 5.15, 5.22(i), 5.31, 6.22, and the whole of Appendix B. Each reader is invited to construct his or her own.

It is a pleasure at this point to acknowledge the help and advice I have had from present and former colleagues: Graham Jameson and Stephen Power at Lancaster both read and made helpful comments on the earlier chapters, and Bob Fraga in Dhahran gave essential technical help with the preparation of the typescript. For my wife June, whose constant encouragement and heroic typing was necessary (and sufficient) to ensure completion of this book, no words of appreciation can be adequate. And finally the staff of John Wiley at Chichester have been both patient and helpful; their assistance too is gratefully acknowledged.

CHAPTER 1

Fourier Coefficients and Fourier Series

1.1 INTRODUCTION

Trigonometric series were considered by mathematicians firstly in response to problems in applied mathematics, of which the two most important are the problem of the vibrating string and the problem of the conduction of heat. The name of Joseph Fourier is associated with the second of these, through his famous book, *La Theorie Analytique de la Chaleur* ('The Analytical Theory of Heat') which was published in 1822. In this introductory paragraph we shall outline a typical but easy problem in the theory of conduction of heat, and its solution by use of trigonometric series, which will in turn motivate our later definitions and results.

Suppose then that we are given an infinite slab of a uniform conducting material, of thickness d (Fig. 1.1), and initial temperature distribution $f_0(x)$, $0 \leqslant x \leqslant d$, which is surrounded by an ambient medium of temperature 0, and that we wish to determine the subsequent temperature distribution $f(x, t)$ for all x and time $t, 0 \leqslant x \leqslant d, t \geqslant 0$.

The physical theory tells us that we have to solve the partial differential equation (the heat equation) which here takes the simplified form

$$\frac{\partial^2 f}{\partial x^2} = k^2 \frac{\partial f}{\partial t} \tag{1.1}$$

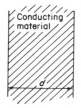

Conducting material

d

Fig. 1.1.

1

where k is a constant depending on the physical properties of the conducting substance, subject to the boundary values

(i) $f(x, 0) = f_0(x)$, and
(ii) $f(0, t) = f(d, t) = 0$, for all t.

(Pure mathematicians may ignore the physical background, and treat this simply as a problem in the theory of partial differential equations.)

To solve the equation, we first find solutions of the form

$$f(x, t) = g(x)h(t), \qquad (1.2)$$

that is, a solution in which the function f is the product of two functions of the individual variables x and t. If we put (1.2) into (1.1) we obtain

$$g''(x)/g(x) = k^2 h'(t)/h(t). \qquad (1.3)$$

Since the two sides of (1.3) involve only the separate variables x and t, both must equal a constant, which we call $-c^2$ (the reason for the minus sign will appear shortly). We then have $g''(x) + c^2 g(x) = 0$, and $h'(t) = -(c/k)^2 h(t)$, a pair of ordinary differential equations, which have the elementary solutions $g(x) = a \cos cx + b \sin cx$, $h(t) = e^{-(c/k)^2 t}$, where the minus sign in the exponent (and $c^2 \geqslant 0$) now ensures that h is bounded for large t. Hence a solution of equation (1.1) of the form (1.2), is given by

$$f(x, t) = e^{-(c/k)^2 t} (a \cos cx + b \sin cx),$$

where a, b are constants, as yet undetermined.

We now consider the effect of the boundary conditions (i) and (ii). (ii) tells us firstly that $f(0, t) = 0$, whence a must be zero, and secondly that $f(d, t) = 0$, whence (since b cannot also be zero), $\sin cd$ must be zero, and hence cd must be a multiple of π: say $c = n\pi/d$, $n = 1, 2, 3, \ldots$. Each value of n gives a new solution, which we write as

$$f_n(x, t) = b_n e^{-(n^2 \pi^2/k^2 d^2)t} \sin(n\pi x/d).$$

We now attempt to satisfy condition (i). Evidently no single solution f_n can do this, but we notice that equation (1.1) is linear in f, and hence we may take sums of the form $f_1 + f_2 + \cdots + f_n + \cdots$ as solutions of (1.1), subject of course to appropriate convergence criteria. This leads us to consider the sum

$$\sum_{n=1}^{\infty} b_n e^{-(n^2 \pi^2/k^2 d^2)t} \sin(n\pi x/d). \qquad (1.4)$$

as a solution of equation (1.1), and the condition (i) requires that

$$f_0(x) = \sum_{n=1}^{\infty} b_n \sin(n\pi x/d). \qquad (1.5)$$

This brings us to the central problems with which we shall have to deal, namely how should we choose the coefficients (in this case the sequence (b_n)), and in what sense does the resulting series represent the initial function f_0.

The first of these is easy and we shall consider it at once: the second is the subject matter of the greater part of this book.

To find the coefficients b_n, we make use of the fact that for integers m, n,

$$\int_0^d \sin(n\pi x/d) \sin(m\pi x/d)\, dx = \begin{cases} 0 & \text{if } m \neq n \\ \tfrac{1}{2}d & \text{if } m = n \geq 1, \end{cases}$$

(see Lemma 1.4 below for this and similar calculations).

Hence if we multiply equation (1.5) throughout by $\sin(m\pi x/d)$ and integrate term by term over the interval $[0, d]$ (and postpone the problem of justifying this process) we obtain

$$\int_0^d f_0(x) \sin(m\pi x/d)\, dx = b_m \tfrac{1}{2} d, \quad \text{or}$$

$$b_n = \frac{2}{d} \int_0^d f_0(x) \sin(n\pi x/d)\, dx. \tag{1.6}$$

Consequently we can state that with this choice of coefficients, the series (1.4) is a solution of equation (1.1) together with boundary conditions (i) and (ii), assuming that our theory allows for a representation of the initial function f_0 in the form (1.5).

For instance, if $f_0(x) = 1$ for $0 < x < d$, then

$$b_n = \frac{2}{d} \int_0^d \sin(n\pi x/d)\, dx = \frac{2}{n\pi}[-\cos(n\pi x/d)]_0^d$$

$$= \frac{2}{n\pi}(1 - \cos n\pi) = \begin{cases} 0 & \text{if } n \text{ is even,} \\ \dfrac{4}{n\pi} & \text{if } n \text{ is odd.} \end{cases}$$

Hence on writing $2r + 1$ for $n, r = 0, 1, 2, \ldots$, we obtain from (1.5) the equation

$$1 = \frac{4}{\pi} \sum_{r=0}^{\infty} \frac{1}{2r+1} \sin((2r+1)\pi x/d), \qquad 0 < x < d,$$

and from (1.4) the formal solution

$$f(x, t) = \frac{4}{\pi} \sum_{r=0}^{\infty} \frac{1}{2r+1} e^{-(2r+1)^2 \pi^2 t / k^2 d^2} \sin((2r+1)\pi x/d)$$

to our original differential equation.

We shall see later how these and similar calculations can be justified; for the moment we leave this preliminary investigation, and begin a systematic development of the theory.

1.2 FC-FUNCTIONS AND THEIR FOURIER COEFFICIENTS

We begin by specifying exactly the class of functions whose Fourier series will be investigated in chapters one to four. These we shall call 'FC-functions' (the mnemonic is 'Finitely Continuous' though the description is not exact): the term conveys that the function is continuous except for a finite (possibly empty) set of discontinuities, near which it remains bounded. More precisely we have the following definition.

Definition 1.1 Let f be a real- or complex-valued function defined on an interval $[a, b]$, a, b being real numbers. Then f is said to be an *FC-function* if (and only if)

  (i) it is bounded: for some $M > 0$, and all x in $[a, b]$, $|f(x)| \leqslant M$, and

  (ii) it is continuous, with at most finitely many exceptions: for some finite set $\{x_0, x_1, \ldots, x_n\}$ with $a = x_0 < x_1 < x_2 < \cdots < x_n = b$, f is continuous on each open interval $(x_0, x_1), (x_1, x_2), \ldots, (x_{n-1}, x_n)$.

The following are some typical examples which will clarify the definition.

Examples 1.2 (i) Any function which is continuous on the *whole* of $[a, b]$ is an FC-function (the boundedness follows from a theorem of elementary analysis). Thus polynomials (in x), sums of sines and cosines, and their moduli are all FC-functions.

  (ii) $f(x) = \begin{cases} |x| & \text{on } [-1, 1] \\ 0 & \text{for other real values of } x \end{cases}$ (Fig. 1.2).

This function is an FC-function on any interval $[a, b]$ – for instance $[-2, 2]$ –

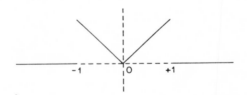

Fig. 1.2.

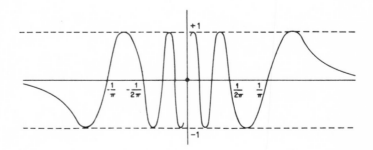

Fig. 1.3.

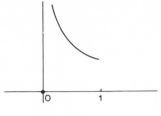

Fig. 1.4.

being bounded and discontinuous only at ± 1.

(iii) $f(x) = \begin{cases} \sin(1/x), & x \neq 0 \\ 0, & x = 0 \end{cases}$ (Fig. 1.3).

This is also an FC-function on any $[a, b]$, since it is continuous except at $x = 0$, and satisfies $|f(x)| \leqslant 1$ for all x.

(iv) $f(x) = \begin{cases} x^{-\alpha}, & 0 < x \leqslant 1 \\ 0, & x = 0 \end{cases}$,   ($\alpha$ a positive constant)   (Fig.1.4).

This is not an FC-function on $[0, 1]$ since $f(x) \to \infty$ as $x \to 0_+$ and so f is not bounded near 0.

(v) $f(x) = \begin{cases} 1, x \text{ rational} \\ 0, x \text{ irrational} \end{cases}$

This function is discontinuous everywhere, and is thus not an FC-function on any interval.

Note We could obtain a slightly more inclusive theory by considering bounded functions with a denumerable set of discontinuities (the class of DC-functions). This class is the uniform closure of our class of FC-functions. However, there would be little practical gain at our level, while for more advanced students the class of Riemann-integrable functions (which are bounded, and continuous except on null sets), or indeed the Lebesgue class L^1 are more suitable.

We know from the elementary theory of integration, or from Appendix A, that the integral $\int_a^b f(x)\,dx$ is defined for FC-functions f, and has the properties of linearity, monotonicity, etc.

For these functions, we have a notion of orthogonality which generalizes the notion of orthogonality for finite-dimensional vectors.

Definition 1.3 Let f, g be FC-functions on $[a, b]$. We say that f, g are *orthogonal* on $[a, b]$ if $\int_a^b f(x)\overline{g(x)}\,dx = 0$.

(For real-valued functions this reduces to $\int_a^b f(x)g(x)\,dx = 0$.)

In general we shall write (f, g) in place of $\int_a^b f(x)\overline{g(x)}\,dx$ where the interval $[a, b]$ is clear from the context. (f, g) is called the *inner product* of the functions

f and g, and generalizes the scalar product $u \cdot v$ of ordinary finite-dimensional vectors: such vectors are of course orthogonal (perpendicular) if $u \cdot v = 0$. Notice that $(f, f) = \int_a^b |f(x)|^2 \, dx$.

(The potential conflict of notation between (a, b) – the open interval whose end points are a and b, and (f, g) – the inner product of functions f and g, should not cause confusion since it will be clear from the context whether the entries are real numbers or functions.)

A sequence $(f_n)_{n=1}^{\infty}$ or $(f_n)_{n=-\infty}^{\infty}$ of functions is said to be *orthogonal* if $(f_m, f_n) = 0$ whenever $m \neq n$. It is said to be *orthonormal* if in addition $(f_n, f_n) = 1$ for all n.

Our fundamental examples of orthogonal sequences of functions are the trigonometric systems on intervals of the form $[0, \pi]$ and $[0, 2\pi]$, as listed in the following result.

Lemma 1.4 (i) The complex exponential sequence $(f_n)_{n=-\infty}^{\infty}$ given by $f_n(x) = e^{inx}$ is orthogonal on $[0, 2\pi]$. Also $(f_n, f_n) = 2\pi$ for all n.

(ii) The trigonometric system $(1, \cos nx, \sin nx)_{n \geq 1}$ is orthogonal on $[0, 2\pi]$. Also $(f_n, f_n) = \pi$ if $f_n(x) = \cos nx$ or $\sin nx$, $n \geq 1$.

(iii) The cosine and sine systems $(\cos nx)_{n \geq 0}$ and $(\sin nx)_{n \geq 1}$ are orthogonal (separately) on $[0, \pi]$. Also $(f_n, f_n) = \frac{1}{2}\pi$ if $f(x) = \cos nx$ or $\sin nx, n \geq 1$. (These are the so-called *half-range systems*.)

Proof We give (i) and (ii), leaving (iii) as an exercise.

For (i) we have $(f_n, f_m) = \int_0^{2\pi} e^{i(n-m)x} \, dx = 2\pi$ if $m = n$, while if $m \neq n$, $(f_n, f_m) = (1/i(n-m))[e^{i(n-m)x}]_0^{2\pi} = 0$.

For (ii) we use the elementary identities

$$\cos nx \cos mx = \tfrac{1}{2}[\cos(n-m)x + \cos(n+m)x],$$

$$\sin nx \sin mx = \tfrac{1}{2}[\cos(n-m)x - \cos(n+m)x], \quad \text{and}$$

$$\cos nx \sin mx = \tfrac{1}{2}[\sin(m+n)x + \sin(m-n)x].$$

Thus for instance,

$$\int_0^{2\pi} \cos nx \cos mx \, dx = \frac{1}{2} \int_0^{2\pi} [\cos(n-m)x + \cos(n+m)x] \, dx$$

$$= \frac{1}{2} \int_0^{2\pi} (1 + \cos 2nx) \, dx \quad \text{if } m = n \geq 1$$

$$= \frac{1}{2} \left[x + \frac{1}{2n} \sin 2nx \right]_0^{2\pi} = \pi,$$

or, if $m \neq n$,

$$= \frac{1}{2} \left[\frac{1}{n-m} \sin(n-m)x + \frac{1}{n+m} \sin(n+m)x \right]_0^{2\pi}$$

$$= 0;$$

the integral with repeated sines is treated similarly. Also

$$\int_0^{2\pi} \cos nx \sin mx \, dx = \frac{1}{2} \int_0^{2\pi} [\sin (m + n)x + \sin (m - n)x] \, dx,$$

(where the second term vanishes if $m = n$)

$$= \frac{1}{2} \left[\frac{1}{m + n} \cos (m + n)x + \frac{1}{m - n} \cos (m - n)x \right]_0^{2\pi}$$

$$= 0 \text{ in all cases.}$$

For a given f and orthogonal sequence (f_n), we define the orthogonal components of f with respect to the sequence (f_n) to be the numbers $c_n(f)$ given by $c_n(f) = (f, f_n)/(f_n, f_n)$: we shall assume that each f_n is not zero (at some point of continuity) so that $(f_n, f_n) > 0$ for all n. These components are defined in such a way that if f can be expressed as a sum $\sum_n c_n f_n$ and if termwise integration is permitted, we can take the scalar product of both sides with any f_m to obtain $(f, f_m) = c_m(f_m, f_m)$, all other terms vanishing, so that the coefficients c_n must have the given form – compare the argument used in Section 1.1. We shall see later (Theorem 1.13(i)) that these coefficients also have a certain unique minimising property which gives an additional reason for their importance. In the particular case when the sequence (f_n) is one of those described in Lemma 1.4, the coefficients $c_n(f)$ are called the Fourier coefficients of f (the term is occasionally used in the general situation too). Formally we define Fourier coefficients as follows.

Definition 1.5 Let f be an FC-function on $[0, 2\pi]$. The *exponential Fourier coefficients* are the numbers $c_n(f)$, $-\infty < n < \infty$, given by the formula

$$c_n(f) = \frac{1}{2\pi} \int_0^{2\pi} f(x)e^{-inx} \, dx.$$

The *trigonometric Fourier coefficients* are the numbers $a_n(f)$, $n \geq 0$, $b_n(f)$, $n \geq 1$, given by the formulae

$$a_n(f) = \frac{1}{\pi} \int_0^{2\pi} f(x) \cos nx \, dx,$$

$$b_n(f) = \frac{1}{\pi} \int_0^{2\pi} f(x) \sin nx \, dx.$$

Notes (i) For Fourier coefficients with respect to the separate trigonometric systems on $[0, \pi]$ (Lemma 1.4 (iii)), see the discussion of odd and even functions below.

(ii) In our preliminary discussion we defined an orthogonal component with respect to a general orthogonal system by the formula

$$c_n(f) = (f, f_n)/(f_n, f_n).$$

It is easy to check that Definition 1.5 is in accordance with this in all cases except when $f_n(x) = 1$ (or $\cos 0x$) in the trigonometric system, when the general formula gives

$$(f, 1)/(1, 1) = \frac{1}{2\pi} \int_0^{2\pi} f(x)\,dx$$

which is not $a_0(f)$ but $\frac{1}{2}a_0(f)$ (and also $c_0(f)$). The above definition of $a_0(f)$ is made in the interest of keeping to a uniform set of formulae for all a_n, $n \geqslant 0$: in return for this we must remember that the constant term in the expansion of f in the trigonometric systems is $\frac{1}{2}a_0$ instead of a_0.

Now that we have defined some orthogonal systems and the components of a general function with respect to them, we turn to our main problem, which is to reconstruct or recover the function from its components. In accordance with the discussion preceding Definition 1.5, and the results of Section 1.1, we make the following definition.

Definition 1.6 Let (f_n) be an orthogonal system on an interval $[a, b]$, and for a given FC-function f, let

$$c_n = c_n(f) = (f, f_n)/(f_n, f_n)$$

$$= \int_a^b f(x)\overline{f_n(x)}\,dx \bigg/ \int_a^b |f_n(x)|^2\,dx.$$

Then the series $\sum_n c_n f_n$ is called the *orthogonal expansion of f with respect to the system* (f_n), and we write

$$f \sim \sum_n c_n f_n$$

to indicate this dependence. (Notice that as yet we have no information concerning possible convergence of the formal sum, so that we use the neutral symbol $\sim$ for the association between the function and the series, rather than the sign of equality.)

In the case of the exponential and trigonometric systems the sum becomes either

$$f \sim \sum_{n=-\infty}^{\infty} c_n e^{inx}, \quad \text{or} \quad f \sim \tfrac{1}{2}a_0 + \sum_{n=1}^{\infty} (a_n \cos nx + b_n \sin nx),$$

where the coefficients (c_n), (a_n), (b_n) are of course given by Definition 1.5. Notice that we write c_n (or a_n or b_n) for $c_n(f)$, etc., when no confusion can arise: note also the occurrence of the factor $\frac{1}{2}$ in the constant term as described above.

Using the formulae for a_n and b_n we find that for $n \geqslant 1$,

$$a_n \cos nx + b_n \sin nx = \frac{1}{\pi} \int_0^{2\pi} f(t)(\cos nx \cos nt + \sin nx \sin nt)\,dt$$

$$= \frac{1}{\pi} \int_0^{2\pi} f(t) \cos n(x - t) \, dt$$

$$= \frac{1}{\pi} \int_0^{2\pi} f(t) \tfrac{1}{2} (e^{in(x-t)} + e^{-in(x-t)}) \, dt$$

$$= c_n e^{inx} + c_{-n} e^{-inx}.$$

(Also $c_0 = \tfrac{1}{2} a_0$ as we have already noted.)

Thus the exponential and trigonometric series are actually the same series, provided that in the exponential case we regard the terms of degree n and $-n$ as being bracketed together. Both of these series will be referred to as the *Fourier series* of f, and denoted $S(f)$: thus we may write

$$f \sim S(f), \quad \text{or} \quad S(f) = \sum_{n=-\infty}^{\infty} c_n e^{inx}, \quad \text{but not (yet!)} \quad f = S(f).$$

(The term 'Fourier series' may also be applied to the general expansion $\sum_n c_n f_n$ of which the exponential and trigonometric series are special cases.)

The theory of Fourier series is of such fundamental importance in applications because it expresses a general function f which may represent a physical quantity such as temperature, or the position of a vibrating system, in terms of component functions which are *periodic:* i.e. there exists some positive constant k such that for each (f_n) in the orthogonal system,

$$f_n(x + k) = f_n(x).$$

The smallest such k is called the *period* of f: in our examples, $k = 2\pi$. With this in mind we shall adopt the following convention which will hold, unless otherwise stated, throughout Chapters 1 to 4.

All functions under consideration will be assumed to be defined for all real values and 2π-periodic:

$$f(x + 2\pi) = f(x) \quad \text{for all real } x.$$

Thus given a function defined on $[0, 2\pi]$ we shall extend its definition to include all real values by the following procedure:

Given any real number y, there is a unique integer n for which

$$y = x + 2n\pi, \quad 0 \leqslant x < 2\pi,$$

and in this case we define (or redefine) $f(y)$ to be equal to $f(x)$.

This evidently produces a 2π-periodic function which agrees with f on $[0, 2\pi)$, though not necessarily at 2π itself.

For example, if $f(x) = x$ on $[0, 2\pi]$, then the extended function has the graph indicated in Fig. 1.5: in particular, $f(2n\pi) = 0$ for all integers n.

We shall occasionally vary this procedure slightly by supposing that the function is defined originally on $[-\pi, \pi]$, and extended periodically by requiring that $f(y) = f(x)$ when $y = x + 2n\pi$, $-\pi \leqslant x < \pi$.

10

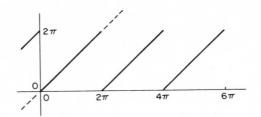

Fig. 1.5.

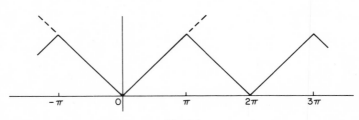

Fig. 1.6.

Which procedure we adopt should be clear from the context. For example, if $f(x) = |x|$ on $[-\pi, \pi]$ we obtain the function in Fig. 1.6.

In each of Figures 1.5 and 1.6, the dashed part of the graph represents the function before modification.

One consequence of the assumption of periodicity is that when integrating a function with period k, we may integrate over *any* interval of length k:

$$\int_0^k f(x)\,dx = \int_a^{a+k} f(x)\,dx \quad \text{for } any \text{ real } a.$$

For instance, we shall often find it more convenient to integrate 2π-periodic functions over $[-\pi, \pi]$ than over $[0, 2\pi]$.

A second consequence of periodicity is that the definition of continuity requires a slight modification. We shall use the phrase 'a continuous k-periodic function' (or sometimes just 'a continuous function') to mean one which remains continuous when extended periodically. Specifically, this requires that the function should be continuous in the usual sense for $0 < x < k$ (for each such x, $f(y) \to f(x)$ as $y \to x$) and also that $f(0)$ and $f(k)$ have a common value for which

$$\lim_{x \to 0_+} f(x) = f(0) = f(k) = \lim_{x \to k_-} f(x).$$

Thus the function in Fig. 1.5 is not continuous in this sense; the function in Fig. 1.6 is continuous, however.

We finish this section with a consideration of the special forms taken by the Fourier series of odd and even functions.

If f is an *even* function ($f(x) = f(-x)$ for all real x), then

$$a_n(f) = \frac{1}{\pi} \int_0^{2\pi} f(x) \cos nx \, dx$$

$$= \frac{1}{\pi} \int_{-\pi}^{\pi} f(x) \cos nx \, dx \qquad \text{(the integrand is periodic)}$$

$$= \frac{1}{\pi} \int_0^{\pi} f(x) \cos nx \, dx + \frac{1}{\pi} \int_0^{\pi} f(-x) \cos nx \, dx$$

(putting $-x$ for x in one half of the integral)

$$= \frac{2}{\pi} \int_0^{\pi} f(x) \cos nx \, dx \qquad \text{(since } f \text{ is even)}.$$

Similarly,

$$b_n(f) = \frac{1}{\pi} \int_0^{\pi} (f(x) - f(-x)) \sin nx \, dx = 0 \text{ if } f \text{ is even.}$$

Thus for an even function,

$$f \sim S(f) = \tfrac{1}{2} a_0 + \sum_{n=1}^{\infty} a_n \cos nx,$$

where

$$a_n = \frac{2}{\pi} \int_0^{\pi} f(x) \cos nx \, dx, \qquad n = 0, 1, 2, \ldots$$

Similar considerations show that if f is *odd* ($f(x) = -f(-x)$ for all real x),

$$f \sim S(f) = \sum_{n=1}^{\infty} b_n \sin nx$$

where

$$b_n = \frac{2}{\pi} \int_0^{\pi} f(x) \sin nx \, dx, \qquad n = 1, 2, 3, \ldots$$

Thus the trigonometric Fourier series of an even function contains only cosine terms, while that of an odd function contains only sine terms. We refer to the exercises at the end of the chapter for the corresponding properties of exponential Fourier series.

Notice finally that a function which is defined initially only on $[0, \pi]$ may be extended to $[-\pi, \pi]$ *either* as an even function (by defining $f(-x) = f(x)$ for $0 \leqslant x \leqslant \pi$) *or* as an odd function (by defining $f(0) = f(\pm \pi) = 0$, $f(-x) = -f(x)$ for $0 < x < \pi$). In the first case its Fourier series will contain only cosine terms, and will correspond to the expansion of f with respect to the half-range system $(\cos nx)_{n \geqslant 0}$ (Lemma 1.4(iii)). In the second case its Fourier series will contain only sine terms and will be the expansion of f with respect to $(\sin nx)_{n \geqslant 1}$. In particular,

12

the expansion

$$1 \sim \frac{4}{\pi} \sum_{r=0}^{\infty} \frac{1}{2r+1} \sin((2r+1)x), \qquad 0 < x < \pi$$

which we obtained at the end of Section 1.1 is of this latter type, since it is simply the Fourier expansion of the function which is $+1$ on $(0, \pi)$, -1 on $(-\pi, 0)$.

1.3 COMPLETENESS

The theory which we outlined in the previous section concerns the relation between a given FC-function f, and its orthogonal expansion $\sum_n c_n f_n$ with respect to a given orthogonal sequence (f_n), the coefficients being given by

$$c_n = c_n(f) = (f, f_n)/(f_n, f_n).$$

It is obviously undesirable in this context that a non-zero function should be associated with the zero series (i.e. $c_n = 0$, all n). However, this is sometimes the case, as we may see by considering a function f which is zero throughout $[a, b]$, except for a finite number of points, where its values may be any real (or complex) number, as indicated in Fig. 1.7.

Such a function f has the property that $\int_a^b f(x)g(x)\,dx = 0$ for any FC-function g, and f is indeed indistinguishable from zero, for purposes of integration at least. In order to avoid this situation we make the following definition.

Definition 1.7 We shall say that a FC-function f is *almost zero* if it is zero except at a finite number of points (equivalently $f(x) = 0$ whenever x is a point of continuity). Two FC-functions f and g will be called *almost equal* if $f - g$ is almost zero.

Note Functions which are almost equal are identical from the point of view of integration. The relation of 'almost equality' is of course an equivalence relation, and we shall 'really' consider not individual functions but their equivalence classes. However, the notational distinctions would be cumbersome and we shall continue to regard individual functions as our basic raw material.

Now that we have defined the notion of almost equality, we can state the property of our orthogonal sequence (f_n) which will make it represent an arbitrary FC-function.

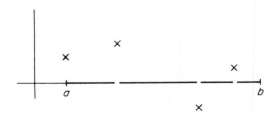

Fig. 1.7.

Definition 1.8 Let (f_n) be an orthogonal sequence of functions on an interval $[a,b]$. We say that (f_n) is *complete on* $[a,b]$ for each FC-function f, $(f,f_n) = \int_a^b f(x)\, f_n(x)\, dx = 0$ for all n, implies that f is almost zero.

An alternative way of stating this definition is to say that for a complete system (f_n) and an FC-function which is non-zero at some point of continuity, (f,f_n) must be non-zero for some n.

We now have to prove the completeness of the basic exponential and trigonometric systems; we need one final definition.

Definition 1.9 A finite linear combination of the functions $(e_n)_{n=-\infty}^{\infty}$ where $e_n(x) = e^{inx}$ is called a *trigonometric polynomial*: $T(x) = \sum_{n=-N}^{N} t_n e^{inx}$, for some complex coefficients (t_n) and $N = 0, 1, 2, \ldots$. The reason for the name appears on putting $z = e^{ix}$ so that $T(x) = \sum_{n=-N}^{N} t_n z^n$. Euler's formulae

$$e^{\pm inx} = \cos nx \pm i \sin nx$$

show that T can be written in the alternative form

$$T(x) = t_0 + \sum_{n=1}^{N} \{(t_n + t_{-n})\cos nx + i(t_n - t_{-n})\sin nx\},$$

and hence that we could equally well define a trigonometric polynomial as a finite sum of cosines and sines.

The expression $T(x) = \sum_{-N}^{N} t_n e^{inx}$ shows that sums and products of trigonometric polynomials are once more trigonometric polynomials – a fact which is vital for the proof which follows.

Theorem 1.10 (i) The exponential system $(e^{inx})_{n=-\infty}^{\infty}$ and the trigonometric system $(1, \cos nx, \sin nx)_{n=1}^{\infty}$ are both complete on $[0, 2\pi]$.

(ii) The cosine and sine systems $(\cos nx)_{n\geqslant 0}$ and $(\sin nx)_{n\geqslant 1}$ are complete on $[0, \pi]$.

Proof (Lebesgue) To prove (i), suppose that f is an FC-function on $[0, 2\pi]$, such that $(f,f_n) = 0$ for all f_n in one or other of the two systems. The preceding discussion shows that in either case this is equivalent to assuming that $(f, T) = 0$ for all trigonometric polynomials T. We now assume that f is *not* almost zero (or equivalently that for some x_0 in $[0, 2\pi]$, f is continuous at x_0, $f(x_0) \neq 0$) and obtain a contradiction.

We may simplify matters, firstly by assuming that f is real valued (for otherwise the hypothesis $(f,f_n) = 0$, with $f_n(x) = \cos nx$ or $\sin nx$, gives the same conclusion for the real and imaginary parts of f), and secondly that $f(x_0) > 0$ (otherwise consider $-f$). Since f is continuous at x_0, there exist positive numbers H ($\frac{1}{2} f(x_0)$ will do) and δ such that $f(x) > H$ for $x_0 - \delta < x < x_0 + \delta$. (We have assumed here that x_0 is not either 0 or 2π: this too entails no loss of generality in the argument since if an FC-function is continuous at 0 with $f(0) > 0$, there are other points near 0 at which f is continuous and positive.) We write I for the interval $(x_0 - \delta,$

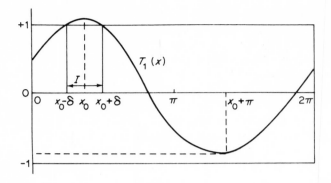

Fig. 1.8.

$x_0 + \delta$) on which $f(x) > H$, and let $I' = [0, 2\pi] \backslash I$, on which the behaviour of f is unknown.

Now consider $T_1(x) = 1 - \cos \delta + \cos(x - x_0)$: this is a trigonometric polynomial which has the property that it is > 1 on I and $\leqslant 1$ on I' (see Fig. 1.8). We also write

$$T_n(x) = (T_1(x))^n, \quad n = 1, 2, 3, \ldots.$$

The discussion following Definition 1.9 shows that $T_n(x)$ is a trigonometric polynomial. We shall show that $\int_I f(x) T_n(x) \, dx \to \infty$ as $n \to \infty$ while $\int_{I'} f(x) T_n(x) \, dx$ remains bounded (a more refined argument would show that in fact $\int_{I'} f(x) T_n(x) \, dx \to 0$).

It will then follow that $\int_0^{2\pi} f(x) T_n(x) \, dx = \int_I + \int_{I'} \to \infty$ as $n \to \infty$, which contradicts the hypothesis that $\int_0^{2\pi} f(x) T_n(x) \, dx = (f, T_n) = 0$.

Consider firstly $\int_I f(x) T_n(x) \, dx$. Both f and T_n are positive on I, and if $|x - x_0| \leqslant \frac{1}{2}\delta$ (note that $\frac{1}{2}$) then $T_n(x) \geqslant c^n$, where we have put $c = 1 - \cos \delta + \cos \frac{1}{2}\delta$, which is > 1.

Hence

$$\int_I f(x) T_n(x) \, dx \geqslant \int_{x_0 - \frac{1}{2}\delta}^{x_0 + \frac{1}{2}\delta} f(x) T_n(x) \, dx \geqslant H \delta c^n \to \infty \text{ as } n \to \infty,$$

as required.

If x is in I' we have $|f(x)| \leqslant M$ for some fixed M (an FC-function is bounded) and $T_n(x) = (T_1(x))^n \leqslant 1$.

Hence

$$\left| \int_{I'} f(x) T_n(x) \, dx \right| \leqslant \int_{I'} |f(x)| |T_n(x)| \, dx$$

$$\leqslant \int_{I'} M \, dx \leqslant 2\pi M.$$

But this is bounded as we asserted previously, and thus the proof of (i) is complete.

The proof of (ii) is a corollary to (i). Suppose for instance that f is an FC-function on $[0, \pi]$ for which $\int_0^\pi f(x) \cos nx \, dx = 0$ for all n.

Define $\tilde{f}$ on $[-\pi, \pi]$ by

$$\tilde{f}(x) = \begin{cases} f(-x) & \text{if } -\pi \leqslant x \leqslant 0 \\ f(x) & \text{if } 0 \leqslant x \leqslant \pi \end{cases}$$

Notice that $\tilde{f}$ is an FC-function on $[-\pi, \pi]$ and that $\int_{-\pi}^\pi \tilde{f}(x) \cos nx \, dx = 2\int_0^\pi f(x) \cos nx \, dx = 0$ by hypothesis, and $\int_{-\pi}^\pi \tilde{f}(x) \sin nx \, dx = 0$ since $\tilde{f}$ is even.

Hence (i) applies, $\tilde{f}$ is almost zero, and thus f is almost zero too.

If $\int_0^\pi f(x) \sin nx \, dx = 0$ for all n, we define $\tilde{f}$ by

$$f(x) = \begin{cases} -f(-x) & \text{if } -\pi < x < 0 \\ 0 & \text{if } x = 0 \text{ or } \pm\pi \\ f(x) & \text{if } 0 < x < \pi \end{cases}$$

and apply a similar argument to this function.

This completes the proof of 1.10.

The theorem which we have just proved is fundamental to the subject of Fourier series. We show its power by deducing our first result on the convergence of Fourier series, and then consider a couple of examples.

Theorem 1.11 Let f be continuous on $[0, 2\pi]$ and suppose that its Fourier series $S(f)$ is uniformly convergent. Then the sum of $S(f)$ is f at each point of $[0, 2\pi]$.

Note The hypothesis of continuity on f implies in particular that $f(0) = f(2\pi)$, as was explained in Section 1.2. The theorem is a hybrid in as much as the hypotheses concern both f and $S(f)$ rather than f or $S(f)$ alone; however, its usefulness will be apparent from the following examples, and the exercises at the end of the chapter.

Proof Write $S_n(f, x) = \sum_{m=-n}^n c_m(f) e^{imx}$ for the nth partial sum of $S(f)$. (This notation will be standard from now on.)

Let g be the sum of $S(f)$: we have to show $f(x) = g(x)$ for all x in $[0, 2\pi]$. Since $S_n \to g$ uniformly and each S_n is continuous, so is g. The uniform convergence of S_n to g also shows that

$$\int_0^{2\pi} S_n(f, x) e^{-ikx} \, dx \to \int_0^{2\pi} g(x) e^{-ikx} \, dx \text{ for any integer } k,$$

that is $c_k(S_n) \to c_k(g)$ as $n \to \infty$

But if $n \geqslant |k|$,

$$c_k(S_n) = \frac{1}{2\pi} \int_0^{2\pi} \left(\sum_{-n}^n c_m(f) e^{imx} \right) e^{-ikx} \, dx = c_k(f)$$

using the orthogonality relations in Lemma 1.4. Combining these results we see that $c_k(f) = c_k(g)$ for all k. This is equivalent to saying that $S(f) = S(g)$, or

$S(f-g)=0$. The completeness of the orthogonal system now shows that $f - g = 0$ at all points of continuity: since f and g are in fact continuous everywhere, this concludes the proof.

Examples 1.12   (i) Let $f(x)=|x|$ on $[-\pi,\pi]$ (f is then defined elsewhere by periodicity). The graph of this function is shown in Fig. 1.6.

This function is even, so the discussion at the end of Section 1.2 shows that $b_n(f)=0$ for all n, and

$$a_n(f) = \frac{2}{\pi} \int_0^\pi f(x) \cos nx \, dx$$

$$= \frac{2}{\pi} \int_0^\pi x \cos nx \, dx.$$

If $n=0$,

$$a_0 = \frac{2}{\pi} \int_0^\pi x \, dx = \frac{2}{\pi} \cdot \frac{1}{2} \pi^2 = \pi, \quad \text{while for } n \geq 1,$$

$$a_n = \frac{2}{\pi} \int_0^\pi x \cos nx \, dx$$

$$= \frac{2}{n\pi} [x \sin nx]_0^\pi - \frac{2}{n\pi} \int_0^\pi \sin nx \, dx$$

$$= 0 + \frac{2}{n^2 \pi} [\cos nx]_0^\pi$$

$$= -\frac{2}{n^2 \pi} (1 - (-1)^n) = \begin{cases} -\dfrac{4}{n^2 \pi} & \text{if } n \text{ is odd,} \\ 0 & \text{if } n \text{ is even.} \end{cases}$$

It follows that

$$S(f) = \tfrac{1}{2} a_0 + \sum_{n=1}^\infty \cos nx$$

$$= \tfrac{1}{2}\pi - \left(\cos x + \frac{1}{3^2} \cos 3x + \frac{1}{5^2} \cos 5x + \cdots \right).$$

Each term of $S(f)$ is in modulus less than or equal to the corresponding term of the convergent series

$$\frac{1}{2}\pi + \frac{4}{\pi} \left(1 + \frac{1}{3^2} + \frac{1}{5^2} + \cdots \right).$$

Hence the Weierstrass M-test (see Theorem A.23 in Appendix A) shows that $S(f)$ is uniformly convergent, and since f is continuous, Theorem 1.11 shows that the sum of $S(f)$ is f, i.e.

$$|x| = \tfrac{1}{2}\pi - \frac{4}{\pi} \sum_{r=0}^{\infty} \frac{1}{(2r+1)^2} \cos(2r+1)x, \qquad -\pi \leqslant x \leqslant \pi.$$

In particular, if we put $x = 0$, we obtain

$$0 = \tfrac{1}{2}\pi - \frac{4}{\pi} \sum_{0}^{\infty} \frac{1}{(2r+1)^2}, \qquad \text{or} \qquad \sum_{0}^{\infty} \frac{1}{(2r+1)^2} = \frac{\pi^2}{8},$$

a famous series whose sum was first found by Euler.

The value of $\sum_{n=1}^{\infty} n^{-2} = s$, say, may be deduced by the well-known device of taking

$$s - \tfrac{1}{4}s = \sum_{n=1}^{\infty} n^{-2} - \sum_{n=1}^{\infty} (2n)^{-2} = 1 + \frac{1}{3^2} + \frac{1}{5^2} + \cdots = \frac{\pi^2}{8}.$$

Hence

$$s = \frac{4}{3}\frac{\pi^2}{8} = \frac{\pi^2}{6}.$$

(ii) Let $f(x) = \cos ax$ on $[-\pi, \pi]$, where a is a real or complex parameter. For the sake of simplicity we shall suppose that a is not an integer (if $a = n$, an integer, then its Fourier series is simply the single term $\cos nx$).

This function is even as in (i), so that $b_n(f) = 0$ for all n, and

$$a_n(f) = \frac{2}{\pi} \int_0^{\pi} \cos ax \cos nx \, dx, \qquad \text{for } n = 0, 1, 2, \ldots.$$

Then

$$a_n(f) = \frac{1}{\pi} \int_0^{\pi} (\cos(a-n)x + \cos(a+n)x) \, dx$$

$$= \frac{1}{\pi} \left[\frac{\sin(a-n)x}{a-n} + \frac{\sin(a+n)x}{a+n} \right]_0^{\pi}$$

(the restriction that a is not an integer ensures that no $a \pm n = 0$)

$$= \frac{1}{\pi} \left(\frac{\sin(a-n)\pi}{a-n} + \frac{\sin(a+n)\pi}{a+n} \right)$$

$$= \frac{(-1)^n 2a \sin a\pi}{\pi(a^2 - n^2)}.$$

Hence

$$S(f) = \tfrac{1}{2}a_0 + \sum_{1}^{\infty} a_n \cos nx$$

$$= \frac{1}{\pi} \sin a\pi \left[\frac{1}{a} + 2a \sum_{n=1}^{\infty} \frac{(-1)^n}{a^2 - n^2} \cos nx \right].$$

This series is absolutely and uniformly convergent by the Weierstrass M-test, and f is continuous, so that from Theorem 1.11 we may deduce that for any complex non-integer value of a, we have

$$\cos ax = \frac{1}{\pi} \sin a\pi \left[\frac{1}{a} + 2a \sum_{n=1}^{\infty} \frac{(-1)^n}{a^2 - n^2} \cos nx \right], \qquad -\pi \leqslant x \leqslant \pi.$$

In particular, putting $x = 0, \pi$ we obtain respectively

$$\frac{\pi}{\sin a\pi} = \frac{1}{a} + 2a \sum_{n=1}^{\infty} \frac{(-1)^n}{a^2 - n^2},$$

and

$$\frac{\pi}{\tan a\pi} = \frac{1}{a} + 2a \sum_{n=1}^{\infty} \frac{1}{a^2 - n^2},$$

results which are usually obtained by the methods of complex analysis.

1.4 MEAN-SQUARE APPROXIMATION

This section deals with an alternative way of regarding Fourier coefficients by means of best mean-square approximation, which gives a further reason for their importance. Given two FC-functions on an interval $[a, b]$, we may measure their 'distance' from one another in a number of ways. In problems concerning uniform convergence, as in Section 1.3, it is natural to take

$$d_0(f, g) = \sup \{ |f(x) - g(x)|; a \leqslant x \leqslant b \}.$$

On the other hand, using the integrability of f and g, it is often more useful to consider

$$d_1(f, g) = \int_a^b |f(x) - g(x)| \, \mathrm{d}x,$$

or

$$d_2(f, g) = \left\{ \int_a^b |f(x) - g(x)|^2 \, \mathrm{d}x \right\}^{1/2}.$$

The maximum separation of f from g at individual points is measured by d_0, while d_1 and d_2 concern average separations. The distance d_2 is particularly interesting in the light of our definition of the inner product (Definition 1.3) since

$$(d_2(f, g))^2 = (f - g, f - g).$$

Note The reader with a knowledge of normed linear spaces will recognize three commonly used norms:

$\qquad d_0(f, g) = \| f - g \|_\infty$, the uniform norm,

$\qquad d_1(f, g) = \| f - g \|_1$, the mean absolute value (or L') norm,

and

$\qquad d_2(f, g) = \| f - g \|_2$, the mean-square norm.

We shall show that among all trigonometric polynomials of degree at most N (Definition 1.9), the Nth partial sum of the Fourier series gives the unique best approximation relative to d_2. More precisely we have the following result.

Theorem 1.13 (i) Let f be an FC-function on $[0, 2\pi]$, and $S(f) = \sum_{-\infty}^{\infty} c_n e^{inx}$ its Fourier series. For any $N = 0, 1, 2, \ldots$, let $S_N(x) = \sum_{-N}^{N} c_n e^{inx}$, and for any complex numbers (t_n), let $T_N(x) = \sum_{-N}^{N} t_n e^{inx}$.
Then

$$\frac{1}{2\pi} \int_0^{2\pi} |f(x) - T_N(x)|^2 \, dx = \frac{1}{2\pi} \int_0^{2\pi} |f(x) - S_N(x)|^2 \, dx + \sum_{-N}^{N} |c_n - t_n|^2.$$

In particular, $d_2(f, T_N)$ is uniquely minimized when $T_N = S_N$.
   (ii) The corresponding result for real series is that if $S_N(f) = \frac{1}{2} a_0 + \sum_1^N (a_n \cos nx + b_n \sin nx)$, $T_N = \frac{1}{2} a'_0 + \sum_1^N (a'_n \cos nx + b'_n \sin nx)$, then

$$\frac{1}{\pi} \int_0^{2\pi} |(f(x) - T_N(x)|^2 \, dx = \frac{1}{\pi} \int_0^{2\pi} |f(x) - S_N(x)|^2 \, dx$$

$$+ \frac{1}{2} |a_0 - a'_0|^2 + \sum_1^N (|a_n - a'_n|^2 + |b_n - b'_n|^2).$$

Proof (i) We use the notation $(f, g) = \int_0^{2\pi} f(x)\overline{g(x)} dx$: as noted above,

$$\int_0^{2\pi} |(x) - g(x)|^2 \, dx = (f - g, f - g)$$

$$= (f, f) - (g, f) - (f, g) + (g, g).$$

In particular if $e_n(x) = e^{inx}$, then

$$T_N(x) = \sum_{-N}^{N} t_n e_n(x), \quad \text{and} \quad c_n = \frac{1}{2\pi}(f, e_n).$$

Hence

$$\int_0^{2\pi} |f(x) - T_N(x)|^2 \, dx = \int_0^{2\pi} |f(x)|^2 \, dx - (f, T_N)$$

$$- (T_N, f) + (T_N, T_N),$$

and

$$(f, T_N) = \left(f, \sum_{-N}^{N} t_n e_n \right)$$

$$= \sum_{-N}^{N} \bar{t}_n (f, e_n) = 2\pi \sum_{-N}^{N} c_n \bar{t}_n,$$

while

$$(T_N, T_N) = 2\pi \sum_{-N}^{N} |t_n|^2.$$

It follows that

$$\int_0^{2\pi} |f(x) - T_N(x)|^2 \, dx = \int_0^{2\pi} |f(x)|^2 \, dx - 2\pi \sum_{-N}^{N} c_n \bar{t}$$

$$- 2\pi \sum_{-N}^{N} \bar{c}_n t_n + 2\pi \sum_{-N}^{N} |t_n|^2$$

$$= \int_0^{2\pi} |f(x)|^2 \, dx - 2\pi \sum_{-N}^{N} |c_n|^2 + 2\pi \sum_{-N}^{N} |c_n - t_n|^2.$$

In particular, if $t_n = c_n$ for all n,

$$\int_0^{2\pi} |f(x) - S_n(x)|^2 \, dx = \int_0^{2\pi} |f(x)|^2 \, dx - 2\pi \sum_{-N}^{N} |c_n|^2,$$

and the result follows from these two equations.

The final statement of (i) follows since we have

$$\sum_{-N}^{N} |c_n - t_n|^2 > 0 \quad \text{unless } c_n = t_n \text{ for all } n.$$

The proof of (ii) uses the orthogonality of the trigonometric system $(1, \cos nx, \sin x)_{n \geq 1}$ in the same way, and is left as an exercise.

Theorem 1.13 has a number of interesting and important corollaries.

Corollaries 1.14 Let f be an FC-function on $[0, 2\pi]$, and $S(f) = \sum_{-\infty}^{\infty} c_n e^{inx}$ its Fourier series. Then

(i) $\dfrac{1}{2\pi} \displaystyle\int_0^{2\pi} |f(x)|^2 \, dx = \dfrac{1}{2\pi} \int_0^{2\pi} |f(x) - S_N(x)|^2 \, dx + \sum_{-N}^{N} |c_n|^2;$

(ii) $\displaystyle\sum_{-\infty}^{\infty} |c_n|^2 \leq \dfrac{1}{2\pi} \int_0^{\pi} |f(x)|^2 \, dx,$

with equality if and only if

$$\int_0^{2\pi} |f(x) - S_N(x)|^2 \, dx \to 0 \quad \text{as} \quad N \to \infty$$

(we shall see in the next section that this is always the case for FC-functions); and
(iii) $c_n(f) \to 0$ as n tends either to $+\infty$, or to $-\infty$.

Note The result in (ii) is known as *Bessel's inequality* (*Bessel's equation* when we have established equality in the next section). The result of (iii), that Fourier coefficients tend to 0 as $|n| \to \infty$, is known as the *Riemann–Lebesgue lemma*. Corresponding results for real coefficients (a_n), (b_n) are left as an exercise.

Proof (i) is a special case of the previous result, Theorem 1.13(i), and is obtained by putting all $t_n = 0$.

For (ii) notice that $\int_0^{2\pi} |f(x) - S_N(x)|^2 \, dx \geqslant 0$, so that for any N,

$$\sum_{-N}^{N} |c_n|^2 \leqslant \frac{1}{2\pi} \int_0^{2\pi} |f(x)|^2 \, dx.$$

The result follows on letting $N \to \infty$.

Since the series $\sum_{-\infty}^{\infty} |c_n|^2$ is convergent, its terms must tend to zero as $|n| \to \infty$, which gives us (iii).

The importance of the Reimann–Lebesgue lemma in integration theory makes it worth while giving the following alternative proof which will be needed in the next chapter. Recall the notation (see Definition A.16 of Appendix A)

$$\omega_f(h) = \int_0^{2\pi} |f(x+h) - f(x)| \, dx$$

for the integral modulus of continuity of a function f, here assumed 2π-periodic.

Lemma 1.15 Let f be an FC-function and $c_n(f)$ its nth Fourier coefficient. Then

$$|c_n(f)| \leqslant \frac{1}{4\pi} \omega_f \left(\frac{\pi}{|n|} \right), \qquad n = 1, 2, 3, \dots.$$

In particular, since $\omega_f(h) \to 0$ as $h \to 0$, $c_n(f) \to 0$ as $|n| \to \infty$.

(The usefulness of this result is that it gives us an estimate for the *rate* at which $c_n \to 0$).

Proof By Definition 1.5,

$$c_n(f) = \frac{1}{2\pi} \int_0^{2\pi} f(x) e^{-inx} \, dx.$$

We now put $x + \pi/|n|$ for x, and obtain

$$c_n(f) = \frac{1}{2\pi} \int_{-\pi/|n|}^{2\pi - \pi/|n|} f(x + \pi/|n|) e^{-inx - in\pi/|n|} \, dx$$

But $e^{-in\pi/|n|} = e^{\pm i\pi} = -1$, so

$$c_n(f) = -\frac{1}{2\pi} \int_{-\pi/|n|}^{2\pi - \pi/|n|} f(x + \pi/|n|) e^{-inx} \, dx$$

$$= -\frac{1}{2\pi} \int_0^{2\pi} f(x + \pi/|n|) e^{-inx} \, dx$$

using the periodicity of f.

Taking the average of these values we obtain

$$c_n(f) = -\frac{1}{2} \cdot \frac{1}{2\pi} \int_0^{2\pi} (f(x + \pi/|n|) - f(x)) e^{-inx} \, dx.$$

Hence

$$|c_n(f)| \leqslant \frac{1}{2} \cdot \frac{1}{2\pi} \int_0^{2\pi} |f(x + (\pi/|n|)) - f(x))| \, dx$$

$$= \frac{1}{4\pi} \omega_f(\pi/|n|), \quad \text{as required.}$$

The interest and indeed the usefulness of the Riemann–Lebesgue lemma lies in the fact that it tells us that an integral of the form $\int_0^{2\pi} f(x)e^{inx} \, dx$ (or the corresponding integral with sin or cos in place of the exponential function) tends to zero as $n \to \infty$, not because the functions themselves tend to zero, but because of the increasing frequency of the oscillatory term which causes cancellation among the areas which make up the integral. This result is the fundamental fact which lies behind the pointwise convergence of Fourier series, to be investigated in the next chapter.

Notice that the Riemann–Lebesgue lemma says that the *terms* of the Fourier series tend to zero as $|n| \to \infty$, a necessary condition for convergence which experience of elementary analysis will have convinced the reader is by no means sufficient. See also exercise 7 at the end of this chapter.

1.5 CONVOLUTION

In this final section of Chapter 1, we introduce the notion of convolution as a form of generalized multiplication for FC-functions. The importance of the concept lies essentially in the fact that given two FC-functions f and g, and the convolution product which we shall denote $f*g$, the Fourier coefficients of $f*g$ are the products of the Fourier coefficients of f and g:

$$c_n(f*g) = c_n(f) \cdot c_n(g).$$

(The results concerning Fourier coefficients are neater for (c_n) than for the real coefficients (a_n), (b_n), so for this section we deal exclusively with the sequence (c_n).)

We begin with the formal definition.

Definition 1.16 Let f and g be 2π-periodic FC-functions. For any real x, let

$$f*g(x) = \frac{1}{2\pi} \int_0^{2\pi} f(t)g(x - t) \, dt.$$

Then $f*g$ is a continuous 2π-periodic function called the *convolution of* f and g.

The definition embodies an assertion about $f*g$ which we shall prove in Lemma 1.18 below. However, first consider a couple of examples.

Notice that $g(x - t)$ is a translation of $g(x)$ (i.e. as shift of the graph of g through an amount t) and hence $f*g$ is a kind of average of translations of g, using f as a weight function: this also holds in reverse, since we shall shortly show that $f*g = g*f$. The factor $1/2\pi$ is introduced so that $1* = 1$ (where 1 denotes the constant function, value 1).

Examples 1.17   (i) Recall that $e_n(x) = e^{inx}$.

Then
$$e_m * e_n = \begin{cases} e_n & \text{if } m = n, \\ 0 & \text{if } m \neq n. \end{cases}$$

This is immediate, since

$$e_m * e_n(x) = \frac{1}{2\pi} \int_0^{2\pi} e^{imt} e^{in(x-t)} \, dt$$

$$= \frac{1}{2\pi} e^{inx} \int_0^{2\pi} e^{i(m-n)t} \, dt,$$

and the result follows from the orthogonality relations, Lemma 1.4.

(ii) Let
$$f(x) = \begin{cases} 1, & 0 < x < \pi, \\ 0 & \text{elsewhere on } [0, 2\pi]. \end{cases}$$

We will calculate $f*f$.

Choose a value of x between 0 and π. In the integral which determines $f*f$, namely $(1/2\pi)\int_0^{2\pi} f(t)f(x-t)\,dt$, we have $f(t) = 1$ on $(0, \pi)$, while $f(x-t) = 1$ on $(x - \pi, x)$ (see Fig. 1.9).

Hence

$$f*f(x) = \frac{1}{2\pi} \int_0^x 1 \, dx = \frac{1}{2\pi} x, \qquad \text{for } 0 \leqslant x \leqslant \pi,$$

so $f*f$ is a linear function $[0, \pi]$. Similarly, as x increases from π to 2π, $f*f$ decreases linearly from its value of $\frac{1}{2}$ at π, to 0 at 2π. Hence $f*f$ is the periodic continuation of the function $(1/2\pi)|x|$ on $[-\pi, \pi]$, whose graph (apart from the factor $1/2\pi$) is shown in Fig. 1.6.

Notice that this example illustrates the assertion in Definition 1.16, that $f*g$ is continuous even when f and g have discontinuities (f and g must still be FC-functions, however).

We now list the formal properties of convolution.

Graph of f

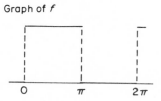
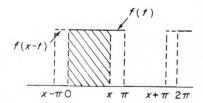

Fig. 1.9.

Lemma 1.18   Let f, g, h be 2π-periodic FC-functions.
  Then (i) $f*g$ is continuous and 2π-periodic.
  (ii) the convolution operations is

  associative: $f*(g*h) = (f*g)*h$,

  commutative: $f*g = g*f$,

and distributive over addition: $f*(g+h) = (f*g)+(f*h)$.
  (iii) $c_n(f*g)$ (the nth Fourier coefficient of $f*g$)

  $= c_n(f)c_n(g)$ (the product of the coefficients of f and g).

Proof   (i) The 2π periodicity of $f*g$ is immediate from that of f and g. Since f is
an FC-function, it must be bounded: $|f(x)| \leqslant M$ for all x, say.
  Then

$$f*g(x) = f*g(x') = \frac{1}{2\pi} \int_0^{2\pi} f(t)g(x-t) - g(x'-t)\,dt,$$

so

$$|f*g(x) - f*g(x')| \leqslant M\frac{1}{2\pi} \int_0^{2\pi} |g(x-t) - g(x'-t)\,dt$$

$$= M\frac{1}{2\pi}\omega_g(|x-x'|),$$

using the notation for the integral modulus of continuity of g.
  Since $\omega_g(\delta) \to 0$ as $\delta \to 0$, it follows that $f*g$ must be continuous.
  (ii) To show commutativity, put $x - t = u$ in the defining integral.
    Thus

$$(f*g)(x) = \frac{1}{2\pi} \int_0^{2\pi} f(t)g(x-t)\,dt$$

$$= \frac{1}{2\pi} \int_x^{x-2\pi} f(x-u)g(u)(-du)$$

$$= \frac{1}{2\pi} \int_{x-2\pi}^{x} f(x-u)g(u)\,du$$

$$= \frac{1}{2\pi} \int_0^{2\pi} f(x-u)g(u)\,du \qquad \text{(by periodicity)}$$

$$= (g*f)(x).$$

To show associativity, we have

$$(f*g)*h(x) = \frac{1}{2\pi} \int_0^{2\pi} (f*g)(t)h(x-t)\,dt$$

$$= \frac{1}{2\pi} \int_0^{2\pi} \left\{ \frac{1}{2\pi} \int_0^{2\pi} f(u)g(t-u)\,du \right\} h(x-t)\,dt$$

$$= \frac{1}{2\pi} \int_0^{2\pi} f(u) \left\{ \frac{1}{2\pi} \int_0^{2\pi} g(t-u)h(x-t)\,dt \right\} du$$

on reversing the order of integration. In the inner integral, put $t' = t - u$ while keeping the limits at 0, 2π (this uses the periodicity of g and h). We get

$$\frac{1}{2\pi} \int_0^{2\pi} f(u) \left\{ \frac{1}{2\pi} \int_0^{2\pi} g(t')h(x-u-t')\,dt' \right\} du$$

$$= \frac{1}{2\pi} \int_0^{2\pi} f(u)(g*h)(x-u)\,du$$

$$= f*(g*h)(x), \text{ as required.}$$

The proof of distributivity over addition is immediate.

(iii) From Definition 1.5 of the nth Fourier coefficient, we have

$$c_n(f*g) = \frac{1}{2\pi} \int_0^{2\pi} (f*g)(x)e^{-inx}\,dx$$

$$= \frac{1}{2\pi} \int_0^{2\pi} \left\{ \frac{1}{2\pi} \int_0^{2\pi} f(t)g(x-t)\,dt \right\} e^{-inx}\,dx$$

$$= \frac{1}{2\pi} \int_0^{2\pi} f(t) \left\{ \frac{1}{2\pi} \int_0^{2\pi} g(x-t)e^{-in(x-t)}\,dx \right\} e^{-int}\,dt$$

on reversing the order of integration

$$= \left\{ \frac{1}{2\pi} \int_0^{2\pi} f(t)e^{-int}\,dt \right\} c_n(g) \qquad \text{using the periodicity of } g$$

$$= c_n(f)c_n(g), \text{ as required.}$$

The results on convolution which we have just proved enable us to complete some of the results on Fourier series from Section 1.3, and obtain some new ones too.

Theorem 1.19 Let f and g be 2π-periodic FC-functions. Then
(i) for all real x,

$$f*g(x) \sum_{n=-\infty}^{\infty} c_n(f)c_n(g)e^{inx},$$

where the series (which is the Fourier series of $f*g$ by Lemma 1.18 (iii)) is absolutely convergent.

(ii) $\dfrac{1}{2\pi} \displaystyle\int_0^{2\pi} f(x)g(-x) = \sum_{-\infty}^{\infty} c_n(f)c_n(g)$ (*Parseval's equation*)

(iii) $\dfrac{1}{2\pi} \displaystyle\int_0^{2\pi} |f(x)|^2\,dx = \sum_{-\infty}^{\infty} |c_n(f)|^2,$ (*Bessel's equation*)

(iv) $\dfrac{1}{2\pi}\displaystyle\int_0^{2\pi} |f(x) - S_n(f, x)|^2 \, dx \to 0$ as ∞

Proof (i) By Lemma 1.18(ii), $f*g$ is continuous and 2π-periodic, and by 1.18(iii), its Fourier series is $\sum_{-\infty}^{\infty} c_n(f)c_n(g)e^{inx}$. If we can show that this series is absolutely convergent, then it will be uniformly convergent by the M-test and the result will follow from Theorem 1.11.

To prove the absolute convergence of $\sum_{-\infty}^{\infty} c_n(f)c_n(g)e^{inx}$, we use the Cauchy–Schwarz inequality (Theorem A.19). We have

$$\sum_{-\infty}^{\infty} |c_n(f)c_n(g)e^{inx}| = \sum_{-\infty}^{\infty} |c_n(f)||c_n(g)|$$

$$\leqslant \left[\sum_{-\infty}^{\infty} |c_n(f)|^2 \sum_{-\infty}^{\infty} |c_n(g)|^2 \right]^{1/2}.$$

But both $\sum |c_n(f)|^2$ and $\sum |c_n(g)|^2$ are convergent by Corollary 1.14 (ii), and the result follows.

If we put $x = 0$ in the result of (i) we obtain

$$f*g(0) = \frac{1}{2\pi} \int_0^{2\pi} f(t)g(-t)\,dt = \sum_{-\infty}^{\infty} c_n(f)c_n(g),$$

which is (ii).

We now take the special case when $g(t) = \overline{f(-t)}$ (where the bar denotes complex conjugation), so that the left-hand side of (ii) is

$$\frac{1}{2\pi} \int_0^{2\pi} |f(x)|^2 \, dx.$$

The Fourier coefficient $c_n(g)$ is found as follows.

$$c_n(g) = \frac{1}{2\pi} \int_0^{2\pi} g(x)e^{-inx}\,dx$$

$$= \frac{1}{2\pi} \int_0^{2\pi} \overline{f(-x)}e^{-inx}\,dx$$

$$= \frac{1}{2\pi} \int_0^{2\pi} \overline{f(x)}e^{inx}\,dx, \qquad \text{putting } -x \text{ for } x$$

$$= \overline{c_n(f)}.$$

Hence the right-hand side of (ii) becomes $\sum_{-\infty}^{\infty} |c_n(f)|^2$, and (iii) follows. Part (iv) is a consequence of (iii) as was noted in the proof of Corollary 1.14.

Note The result of part (iv) may be stated as '$d_2(f, S_n) \to 0$ as $n \to \infty$' (using the notion of distance introduced at the beginning of Section 1.4), or that S_n converges to f in mean square. We shall show in the next chapter that it is *not* generally true that S_n converges to f uniformly (or even pointwise).

Example 1.20 Choose a real number h in the interval $[0, 2\pi]$, and let

$$f(x) = \begin{cases} 1 & \text{if } 0 < x \leqslant h, \\ 0 & \text{if } h < x \leqslant 2\pi. \end{cases}$$

Then

$$c_n(f) = \frac{1}{2\pi} \int_0^{2\pi} f(x) e^{-inx}\, dx = h/2\pi \text{ if } n = 0.$$

If $n \neq 0$, we have

$$c_n(f) = \frac{1}{2\pi} \int_0^h e^{inx}\, dx = -\frac{1}{2\pi in}[e^{-inx}]_0^h$$

$$= \frac{1}{2\pi in}(1 - e^{-inh}) = \frac{1}{\pi n} e^{-inh/2} \sin(\tfrac{1}{2}nh).$$

Then Bessel's equation for this function states that

$$\frac{1}{2\pi} \int_0^{2\pi} |f(x)|^2\, dx = h/2\pi$$

$$= \sum_{-\infty}^{\infty} |c_n(f)|^2$$

$$= |c_0(f)|^2 + \sum_{n \neq 0} (\sin(\tfrac{1}{2}nh)/\pi n)^2$$

$$= (h/2\pi)^2 + 2/\pi^2 \sum_{n=1}^{\infty} \sin^2(\tfrac{1}{2}nh)/n^2.$$

On rearrangement, this gives

$$\sum_{n=1}^{\infty} \sin^2(\tfrac{1}{2}nh)/n^2 = \tfrac{1}{8}h(2\pi - h), \qquad 0 \leqslant h \leqslant \pi.$$

EXERCISES

1 Sketch each of the function given by the following formulae and find its Fourier series in either the exponential or trigonometric form. To which of them does Theorem 1.11 apply? In parts (i) and (iii), h is a real number, $0 < h \leqslant \pi$.

(i) $f(x) = \begin{cases} 1 & \text{for } 0 < x < h, \\ -1 & \text{for } -h < x < 0 \\ 0 & \text{elsewhere on } [-\pi, \pi]. \end{cases}$

(ii) $f(0) = f(2\pi) = 0, \qquad f(x) = \pi - x$ for $0 < x < 2\pi$.

(iii) $f(x) = \begin{cases} 1 - |x|/h & \text{for } 0 \leqslant |x| \leqslant h, \\ 0 & \text{elsewhere on } [-\pi, \pi]. \end{cases}$

(iv) $f(x) = \begin{cases} \sin\frac{1}{2}x & \text{if } -\pi < x < \pi, \\ 0 & \text{if } x = \pm\pi. \end{cases}$

(v) $f(x) = |\sin x|$ for all x.

(vi) $f(x) = x^2$ on $[-\pi, \pi]$.

2   Let f be a 2π-periodic FC-function. Show that f is even if and only if $c_n(f) = c_{-n}(f)$ for all n, and that f is odd if and only if $c_n(f) = -c_{-n}(f)$. Show that f is real-valued if and only if $\overline{c_n(f)} = c_{-n}(f)$ (where the bar denotes complex conjugation).

3   Given two complex sequences $\alpha = (\alpha_n)$, $\beta = (\beta_n)$, we may define their inner product $(\alpha, \beta) = \sum_n \alpha_n \overline{\beta_n}$. This will converge for instance if one of the sequences, say (α_n), is absolutely summable ($\sum_n |\alpha_n|$ convergent), and the other is bounded, or alternatively by the Cauchy–Schwarz inequality if both are square-summable ($\sum_n |\alpha_n|^2$, $\sum_n |\beta_n|^2$ both convergent).
Show that the Parseval's equation (Theorem 1.19(ii)) may be stated in the equivalent form

$$(f, g) = 2\pi(c, d), \text{ where } c = (c_n) = (c_n(f))_{-\infty}^{\infty},$$
$$d = (d_n) = (c_n(g))_{-\infty}^{\infty}$$

4   We have shown in Example 1.12(ii) that if x is not an integer, then

$$\frac{\pi}{\tan \pi x} = \frac{1}{x} + 2x \sum_{n=1}^{\infty} (x^2 - n^2)^{-1}.$$

Deduce by differentiation that

$$\frac{\pi^2}{\sin^2 \pi x} = \sum_{n=-\infty}^{\infty} (x - n)^{-2},$$

and by integration that

$$\sin \pi x = \pi x \prod_{n=1}^{\infty} (1 - x^2/n^2).$$

For real x, the results of Appendix A may be used. For complex x, the results may be deduced as in the real case by the analogous results from complex analysis, or deduced from the real cases by analytic continuation.)

5   Use Bessel's equation (Theorem 1.19(iii)) together with Example 1.12(i) to find the value of $\sum_{r=0}^{\infty} (2r + 1)^{-4}$. Deduce the value of $\sum_{n=1}^{\infty} n^{-4}$.

6   Use Example 1.20 to find the sum of the series

$$\sum_{n=1}^{\infty} \frac{1}{n^2} \cos nh, \quad \text{for } |h| \leqslant \pi.$$

7   Let f be an FC-function on $[a, b]$, and let g be any of the functions $g(x) = e^{ix}$, $\cos x$. Imitate the proof of Lemma 1.15 to show that $I(A) = \int_a^b f(x)g(Ax)\,dx$ tends to zero as $A \to \infty$. (This is the continuous analogue of the Riemann–Lebesgue lemma.)

CHAPTER 2

Convergence Theory

In Chapter 1 we introduced the Fourier series associated with a 2π-periodic function f. In Definition 1.6 we stressed that we had no information concerning possible convergence of the series, and in the remainder of the chapter we proved only a single result (Theorem 1.11) concerning convergence. A number of the examples, notably 1.12 and 1.20, and the exercises to Chapter 1 showed what use could be made of this one theorem: however, a function with even a single discontinuity remained beyond our reach as far as convergence of $S(f)$ is concerned.

In this chapter we shall concentrate on the mathematical problem of the convergence of a Fourier series either at an individual point, or on a subinterval of $[0, 2\pi]$: in the latter cases questions of uniformity must also be considered. We shall impose two alternative restrictions on the function, namely Lipschitz continuity and monotonicity, which correspond in simplified forms to the classical results of Dini, and of Dirichlet and Jordan. We also show at the end of the chapter that continuity alone is not a sufficient condition for convergence of $S(f)$. We shall consider some possible uses of the theory in Chapter 3 and subsequent chapters.

Despite a number of simplifications, Chapter 2 is the most demanding in terms of analytical technique. A reader who is mainly interested in the applications of the theory should at least read Section 2.1 on pointwise convergence, but could initially take the results on uniformity on trust. The Gibbs phenomenon (Section 2.3) and the possibility of divergence of the Fourier series of a continuous function at some points (Section 2.5) are likewise facts which are essential knowledge for applications of the theory, but whose proofs could be omitted at a first reading.

2.1 POINTWISE CONVERGENCE

We begin with the most straightforward of convergence problems: that is, to find a restriction on the behaviour of a function which will ensure the convergence of its Fourier series at a given point. It turns out, perhaps surprisingly, that a smoothness condition only in the neighbourhood of a point x_0, is sufficient to make $S(f)$ convergent. We begin with a simple calculation which puts the partial sums $S_n(f, x)$ into a convenient form. We shall feel free to write $S_n(f)$ or $S_n(x)$ for these sums depending on whether the dependence on f or x is to be emphasized; this was already done in Theorem 1.13.

Lemma 2.1 Let f be an FC-function on $[0, 2\pi]$ and let

$$S_n(f, x) = \sum_{m=-n}^{n} c_m e^{imx} = \tfrac{1}{2}a_0 + \sum_{m=1}^{n} (a_m \cos mx + b_m \sin mx)$$

be the nth partial sum of its Fourier series.

For $n = 0, 1, 2, \ldots$, let $D_n(x) = \begin{cases} \sin(n + \tfrac{1}{2})x/\sin\tfrac{1}{2}x, & 0 < x < 2\pi, \\ 2n + 1, & x = 0, 2\pi. \end{cases}$

Then D_n is a trigonometric polynomial of degree n, and

$$S_n(f, x) = (f * D_n)(x).$$

Proof Recall from Definition 1.6 that we consider *symmetric partial sums* $\sum_{-n}^{n}$ of the complex form of $S(f)$. If we put the formulae for (c_n), the Fourier coefficients of f (Definition 1.5), into the given expression $S_n(f, x) = \sum_{-n}^{n} c_m e^{imx}$, we obtain

$$S_n(f, x) = \sum_{-n}^{n} \left\{ \frac{1}{2\pi} \int_0^{2\pi} f(t)e^{-imt}\, dt \right\} e^{imx}$$

$$= \frac{1}{2\pi} \int_0^{2\pi} f(t) \left\{ \sum_{-n}^{n} e^{im(x-t)} \right\} dt$$

$$= f * D_n(x) \qquad \text{(recall Definition 1.16 of convolution),}$$

where

$$D_n(x) = \sum_{-n}^{n} e^{imx} \text{ is a trigonometric polynomial of degree } n.$$

Thus to complete the lemma, we have to prove the given formula for D_n.

If $\qquad x = 0, 2\pi, D_n(x) = \sum_{-n}^{n} 1 = 2n + 1.$

If $\qquad 0 < x < 2\pi, D_n(x) = e^{-imx} \sum_{r=0}^{2n} e^{irx}, \qquad \text{putting } m = r - n,$

$$= e^{-inx} \frac{e^{i(2n+1)x} - 1}{e^{ix} - 1}$$

using the elementary formula for the sum of a geometric progression This expression simplifies to give

$$\frac{e^{i(n+\frac{1}{2})x} - e^{-i(n+\frac{1}{2})x}}{e^{i\frac{1}{2}x} - e^{-i\frac{1}{2}x}}$$

$$= \sin(n + \tfrac{1}{2})x/\sin(\tfrac{1}{2}x), \text{ as required.}$$

Definition 2.2 The trigonometric polynomial $D_n(x)$ introduced in Lemma 2.1 is called the *Dirichlet kernel*: its graph is indicated in Fig. 2.1. The two forms in which D_n is written in the course of Lemma 2.1 give the estimates

(i) $|D_n(x)| \leqslant 2n + 1$ for all x, and

(ii) $|D_n(x)| \leqslant 1/|\sin\frac{1}{2}x|$ for x not a multiple of 2π.

Notice that D_n may also be written $D_n(x) = 1 + 2\sum_{n=1}^{n} \cos mx$, and hence that

(iii) $\int_0^{2\pi} D_n(x)\,dx = 2\pi$.

Our first pointwise convergence result says that if f is Lipschitz continuous at a point x_0, then $S(f)$ is convergent at $x = x_0$ with sum $f(x_0)$: this is part (i) of Theorem 2.3 below. The proof requires our expression of S_n as $f*D_n$, and also the Reimann–Lebesgue lemma which was proved in Chapter 1 (Corollary 1.14(iii)).

Theorem 2.3 Let f be a 2π-periodic FC-function, and let

$$S_n(f, x) = \sum_{m=-n}^{n} c_m(f)e^{imx}$$

be the nth partial sum of its Fourier series. Then

(i) if f is Lipschitz continuous at a point x_0, then

$$S_n(f, x_0) \to f(x_0) \text{ as } n \to \infty,$$

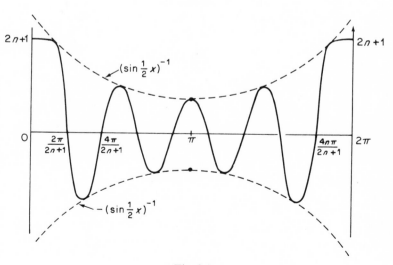

Fig. 2.1.

(ii) if f has right- and left-hand limits at x_0, say l_+ and l_- respectively, such that for some M and all $h > 0$,

$$|f(x_0 + h) - l_+| + |f(x_0 - h) - l_-| \leqslant Mh, \qquad \text{then}$$

$$S_n(f, x_0) \to \tfrac{1}{2}(l_+ + l_-) \text{ as } n \to \infty;$$

(iii) if there is some number c such that for some M and all $h > 0$,

$$|f(x_0 + h) + f(x_0 - h) - 2c| \leqslant Mh, \text{ then}$$

$$S_n(f, x_0) \to c \qquad \text{as } n \to \infty.$$

Proof The hypothesis of (ii) says that f has Lipschitz limits on the right and left at x_0, while the hypothesis in (iii) requires more generally that $f(x_0 + h) + f(x_0 - h)$ has a Lipschitz limit at x_0, so that some cancellation of right- and left-hand values of f near x_0 is allowed. Notice that (i) is a special case of (ii) when $l_+ = l_- = f(x_0)$; while (ii) is a special case of (iii) when $c = \tfrac{1}{2}(l_+ + l_-)$. None the less, we shall prove (i) first for the sake of the clarity of the argument in this case, and we shall then point out how the proof can be modified to give (iii).

Suppose then that f is a 2π-periodic FC-function, which is Lipschitz continuous at x_0, i.e. for some M,

$$|f(x) - f(x_0)| \leqslant M|x - x_0| \quad \text{for all } x \text{ in } [0, 2\pi].$$

From Lemma 2.1 we have

$$S_n(f, x_0) = (f * D_n)(x_0)$$

$$= \frac{1}{2\pi} \int_0^{2\pi} f(x_0 - t) D_n(t) \, dt,$$

and from Definition 2.2 (iii),

$$f(x_0) = \frac{1}{2\pi} \int_0^{2\pi} f(x_0) D_n(t) \, dt.$$

Hence on subtracting we obtain

$$S_n(f, x_0) - f(x_0) = \frac{1}{2\pi} \int_0^{2\pi} \{f(x_0 - t) - f(x_0)\} D_n(t) \, dt$$

$$= \frac{1}{2\pi} \int_0^{2\pi} \frac{f(x_0 - t) - f(x_0)}{\sin \tfrac{1}{2}t} \sin(n + \tfrac{1}{2})t \, dt$$

$$= \frac{1}{2\pi} \int_0^{2\pi} g(x_0, t) \sin(n + \tfrac{1}{2})t \, dt, \qquad \text{say.}$$

Now $g(x_0, t)$, considered as a function of t, is continuous except at the discontinuities of f, and possibly at $0, 2\pi$ where $\sin \tfrac{1}{2}t = 0$. Near $t = 0$ we have

$$|g(x_0, t)| \leqslant \frac{M|t|}{|\sin \tfrac{1}{2}t|},$$

which is bounded as $t \to 0$; thus g is bounded near $t = 0$, and near $t = 2\pi$ by periodicity. Hence g is an FC-function, and on putting

$$\sin(n + \tfrac{1}{2})t = \sin nt \cos \tfrac{1}{2}t + \cos nt \sin \tfrac{1}{2}t$$

we see that $S_n(f, x_0) - f(x_0)$ is a sum of Fourier coefficients of $g(x_0, t) \cos \tfrac{1}{2}t$ or $g(x_0, t) \sin \tfrac{1}{2}t$, and hence must tend to zero by the Riemann–Lebesgue lemma. (Alternatively Exercise 7 of Chapter 1 may be applied directly.) This proves (i).

To prove (iii) we notice that $D_n(t) = D_n(-t)$ and hence that the above integral for S_n may be written

$$S_n(f, x_0) = \frac{1}{2\pi} \int_0^{2\pi} f(x_0 + t) D_n(t) \, dt$$

$$= \frac{1}{2\pi} \int_0^{2\pi} \tfrac{1}{2}(f(x_0 + t) + f(x_0 - t)) D_n(t) \, dt.$$

Hence

$$S_n(f, x_0) - c = \frac{1}{2\pi} \int_0^{2\pi} \left\{ \frac{\tfrac{1}{2}(f(x_0 + t) + f(x_0 - t)) - c}{\sin \tfrac{1}{2}t} \right\} \sin(n + \tfrac{1}{2})t \, dt$$

$$= \frac{1}{2\pi} \int_0^{2\pi} g_1(x_0, t) \sin(n + \tfrac{1}{2})t \, dt, \qquad \text{say.}$$

But the condition (iii) shows as before that g_1 is an FC-function and so the integral tends to zero by the Riemann–Lebesgue lemma.

We illustrate this theorem by some examples.

Examples 2.4 (i) In Example 1.12(i) we found that if $f(x) = |x|$ on $[-\pi, \pi]$, then

$$S(f) = \tfrac{1}{2}\pi - \frac{4}{\pi} \sum_{n=0}^{\infty} (2n + 1)^{-2} \cos(2n + 1)x.$$

The function f is Lipschitz continuous everywhere, so Theorem 2.3(i) gives us an alternative reason for the convergence of $S(f)$ to f, which we had already found in Example 1.12(i).

In this example the convergence is uniform; we shall see in the next section that this always happens when a function is Lipschitz continuous on an interval.

Similar considerations apply to Example 1.12(ii).

(ii) $\qquad\qquad$ Let $f(x) = \begin{cases} \tfrac{1}{2}(\pi - x) & \text{if } 0 < x < 2\pi \\ 0 & \text{if } x = 0, \text{ or } x = 2\pi. \end{cases}$

(this is, apart from the factor $\tfrac{1}{2}$, Exercise 1(ii) from Chapter 1).

f is an odd function, whose graph is shown in Fig. 2.2

The Fourier coefficients are given by

$$a_n = 0 \quad \text{all } n,$$

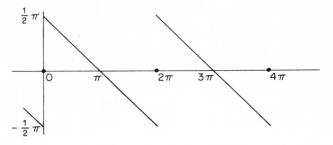

Fig. 2.2.

$$b_n = \frac{2}{\pi}\int_0^\pi \tfrac{1}{2}(\pi - x)\sin nx\,dx, \qquad n = 1, 2, 3, \ldots$$

$$= \frac{1}{n\pi}[-(\pi - x)\cos nx]_0^\pi - \frac{1}{n\pi}\int_0^\pi \cos nx\,dx$$

$$= \frac{1}{n\pi}(\pi - 0) - 0$$

$$= \frac{1}{n}.$$

Hence $S(f) = \sum_{n=1}^{\infty}(1/n)\sin nx$.

This function is Lipschitz continuous on the open interval $(0, 2\pi)$ and Theorem 2.3(i) ensures that $S(f)$ is convergent with sum f on this interval. At $x = 0, 2\pi$, all terms of the series are zero and thus it converges to f at these points too – notice that this is in accordance with Theorem 2.3(ii).

(iii) Let $0 < h \leqslant \pi$, and

$$f(x) = \begin{cases} 1 & \text{on } (0, h), \\ -1 & \text{on } (-h, 0), \\ 0 & \text{elsewhere on } [-\pi, \pi]. \end{cases}$$

(This is Exercise 1(i) from Chapter 1).

The Fourier series here is

$$S(f) = \frac{4}{\pi}\sum_{n=1}^{\infty}\frac{1}{n}\sin^2\tfrac{1}{2}nh \sin nx \qquad \text{(we omit the calculation)}.$$

Theorem 2.3(i) and (ii) show that $S(f)$ is convergent to f at all points except $\pm h$ (when $h < \pi$) when the sum is $\pm\frac{1}{2}$ respectively, instead of 0.

(iv) Let $f(0) = 0$, $f(x) = \sin(\pi^2/x)$ for $0 < |x| < \frac{1}{2}\pi$, and let f be an arbitrary FC-function for $(\pi/2) \leqslant |x| \leqslant \pi$.

Then f is bounded ($|f| \leqslant 1$ for $|x| \leqslant \frac{1}{2}\pi$) and has an oscillatory discontinuity at $x = 0$ (compare Example 1.2(iii)). There is no explicit formula for the Fourier coefficients in this case, but we can see that for $|t| < \frac{1}{2}\pi$, $f(t) + f(-t) = 0$, so that Theorem 2.3(iii) holds with $c = 0$.

Hence $S(f)$ is convergent with sum $f(0) = 0$ at $x = 0$.

2.2 UNIFORM CONVERGENCE

We saw in Section 2.1 that the Fourier series of a function f converges at x_0 with sum $f(x_0)$ at any point at which f is Lipschitz continuous. We shall now show that this extends to the situation in which f is Lipschitz continuous on an interval $I = (a, b)$: we find that convergence is now *uniform* on any subinterval of the form $[a + \delta, b - \delta]$ for positive δ. We could not expect uniform convergence on the whole of the interval I since there may be discontinuities at a or b. The behaviour of $S(f)$ near a point of discontinuity is investigated in Section 2.3.

Recall (see Definition A.5) that f is said to be Lipschitz continuous on an interval $I = (a, b)$ if there exists some positive number M such that

$$|f(x) - f(y)| \leqslant M|x - y|$$

for all x, y in I.

Theorem 2.5 Let f be a 2π-periodic FC-function, and for some interval $I = (a, b) \subseteq [0, 2\pi]$ (or $\subseteq [-\pi, \pi]$ etc.), suppose that f is Lipschitz continuous on I. Then for any δ such that $0 < \delta < \frac{1}{2}(b - a)$, $S(f)$ is uniformly convergent, with sum f on $[a + \delta, b - \delta]$.

Proof We are given that

$$|f(x + h) - f(x)| \leqslant M|h|$$

whenever x and $x + h$ are in $I = (a, b)$.

Let $\delta > 0$ be given and let $I' = [a + \delta, b - \delta]$: we have to show $S_n(f) \to f$ uniformly on I'.

From the proof of Theorem 2.3 we have the formula

$$S_n(f, x) - f(x) = \frac{1}{2\pi} \int_{-\pi}^{+\pi} (f(x - t) - f(x)) D_n(t)\, dt$$

$$= \frac{1}{2\pi} \int_{-\pi}^{\pi} \frac{f(x - t) - f(x)}{\sin \frac{1}{2}t} \sin(n + \tfrac{1}{2})t\, dt.$$

Now suppose that x is in I' and that $\varepsilon > 0$ is given. For any $\delta_1 \leqslant \delta$, write the integral as

$$\frac{1}{2\pi} \int_{-\delta_1}^{\delta_1} \frac{f(x - t) - f(x)}{\sin \frac{1}{2}t} \sin(n + \tfrac{1}{2})t\, dt + J_x$$

say, where J_x is the integral over $[-\pi, \pi] \setminus [-\delta_1, \delta_1]$.

The Lipschitz condition of f shows that if $|t| \leqslant \delta_1$, so that $x - t$ is in I, then

$$\left| \frac{f(x-t)-f(x)}{\sin\frac{1}{2}t} \right| \leqslant \frac{M|t|}{|\sin\frac{1}{2}t|},$$

which is bounded (e.g. by $3M$) as $t \to 0$, since $t/\sin\frac{1}{2}t \to 2$ as $t \to 0$. Using this estimate (and replacing $\sin(n+\frac{1}{2})t$ by 1) we see that the integral over $[-\delta_1, \delta_1]$ is at most

$$\frac{1}{2\pi} 2\delta_1 3M \text{ which can be made } < \varepsilon \text{ if } \delta_1 < \frac{\pi\varepsilon}{2M}.$$

It remains to show that, having fixed this value of δ_1, we can make $|J_x| < \varepsilon$ by making n large enough.

Define

$$g(t) = \begin{cases} 0 & \text{if } |t| < \delta_1, \\ 1/\sin\frac{1}{2}t & \text{if } \delta_1 \leqslant |t| \leqslant \pi: \end{cases}$$

g is an FC-function and we may write

$$J_x = \frac{1}{2\pi} \int_{-\pi}^{\pi} (f(t-t)-f(x))g(t)\sin(n+\tfrac{1}{2})t\,dt.$$

As in the proof of Theorem 2.3 this shows that J_x is a sum of Fourier coefficients of $(f(x-t)-f(x))g(t)$ (as a function of t) and hence by Lemma 1.15 it is sufficient to show that the integral modulus of continuity, $\omega(h)$ say, of this function tends to zero (uniformly with respect to x) as $h \to 0$.

But

$$\omega(h) = \int_{-\pi}^{\pi} |(f(x-t-h)-f(x))g(t+h)=(f(x-t)-f(x))g(t)|\,dt$$

$$= \int_{-\pi}^{\pi} |(f(x-t-h)-f(x-t))g(t+h)+(f(x-t)-f(x))(g(t+h)-g(t))|\,dt$$

$$\leqslant M_1\omega_f(h) + 2M_2\omega_g(h)$$

where M_1, M_2 are bounds for $|g|, |f|$ respectively on $[-\pi, \pi]$. This estimate is independent of x, and $\omega_g(h)$, $\omega_g(h)$ both tend to zero as $h \to 0$, and the proof of Theorem 2.5 is complete.

Example 2.6 In Exercise 2.4(ii) we saw that if $f(x) = \frac{1}{2}(\pi - x)$ on $(0, 2\pi)$, $f(0) = f(2\pi) = 0$, then $S(f)$ is convergent to f at all points. The fact that f is Lipschitz continuous on $(0, 2\pi)$,

$$|f(x) - f(y)| \leqslant \tfrac{1}{2}|x-y| \qquad \text{for } x, y \text{ in } (0, 2\pi),$$

means that we can apply Theorem 2.5 to show that the convergence is uniform on any interval of the form $[\delta, 2\pi - \delta]$, with $0 < \delta < \pi$.

We should not expect uniform convergence on $(0, 2\pi)$ in view of the discontinuities at the end points.

For an alternative argument leading to the uniform convergence of certain special series (of which this is an example) see exercise 11 at the end of this chapter.

It is an important corollary of Theorem 2.5 that the convergence of $S(f)$ is determined only by the local behaviour of f: more precisely, if two functions agree on an open interval then their Fourier series converge or diverge together (and uniformly on closed subintervals). This is the *Riemann localization principle*:

Corollary 2.7 Let f and g be 2π-periodic FC-functions, such that $f(x) = g(x)$ for x in some interval (a, b). Then

$$S_n(f, x) - S_n(g, x) \to 0$$

uniformly for x in any interval of the form $[a + \delta, b - \delta]$ with $0 < \delta < \frac{1}{2}(b - a)$.

Proof This is an immediate corollary of Theorem 2.5 since

$$S_n(f, x) - S_n(g, x) = S_n(f - g, x) \text{ and } (f - g)(x) = 0 \text{ for } x \text{ in } I.$$

2.3 BEHAVIOUR NEAR DISCONTINUITIES

The first two sections of this chapter have shown that if a function has a reasonable degree of smoothness (Lipschitz continuity) at a point or on an interval then its Fourier series will converge with f as its sum, at the point or uniformly on the interval, in the respective cases. We now turn to the problem of the behaviour of the series near a simple jump discontinuity – that is a point at which left- and right-hand limits exist but are unequal. Theorem 2.3(ii) shows in this case that under suitable hypotheses the series will converge to the average of the left- and right-hand limits: in this section we shall show that the partial sums (which being continuous, cannot converge uniformly by Theorem A.25(i)) have local maxima which in fact overshoot the amount of the discontinuity by a fixed proportion – roughly 18%. We begin by considering an example, and then show that the general case follows from it.

Lemma 2.8 Let

$$S_n(x) = \sin x + \tfrac{1}{3}\sin 3x + \tfrac{1}{5}\sin 5x + \cdots$$

$$+ \frac{1}{2n + 1}\sin(2n + 1)x, \qquad \text{for } n = 0, 1, 2, \ldots.$$

Then

(i) $|S_n(x)| < \frac{1}{2}\pi$ for all n and all real x, and

(ii) $\sup\{|S_n(x)|: 0 \leqslant x \leqslant 2\pi\} \to \dfrac{1}{2}\displaystyle\int_0^\pi \dfrac{\sin x}{x}\,dx \quad \text{as } n \to \infty.$

Proof Notice that S_n is $\pi/4$ times the partial sum of the Fourier series of the function which is equal to $+1$ on $(0, \pi)$ and -1 on $(-\pi, 0)$ – this is the case $h = \pi$ of Example 2.4(iii). Our first conclusion shows that these partial sums are uniformly bounded, and the second conclusion gives the exact limit of the maximum value which $|S_n(x)|$ attains on $[0, 2\pi]$ as $n \to \infty$.

Since $S_n(-x) = -S_n(x)$ we need consider only the behaviour of S_n on $[0, \pi]$: since also $S_n(\pi - x) = S_n(x)$ we can further restrict ourselves to the interval $[0, \frac{1}{2}\pi]$.

On differentiating we find that

$$S'_n(x) = \cos x + \cos 3x + \cos 5x + \cdots + \cos(2n + 1)x$$

$$= \frac{1}{2} \sum_{m=0}^{n} e^{i(2m+1)x} + e^{-i(2m+1)x}$$

$$= \frac{1}{2} e^{-i(2n+1)x} \sum_{r=0}^{2n+1} e^{i2rx}$$

$$= \frac{1}{2} e^{-i(2n+1)x} \left\{ \frac{e^{i(4n+4)x} - 1}{e^{i2x} - 1} \right\}$$

$$= \frac{1}{2} \sin 2(n+1)x / \sin x, \text{ if } x \text{ is not a multiple of } \pi.$$

We also have $S'_n(0) = n + 1$, $S'_n(\pi/2) = 0$, $S'_n(\pi) = -(n+1)$, and $S'_n(x) = 0$ when $x = k\pi/2(n+1)$, where $k = 1, 2, \ldots, n$ gives the values of x in $(0, \pi/2)$.

Hence

$$S_n(x) = \frac{1}{2} \int_0^x \frac{\sin 2(n+1)t}{\sin t} dt,$$

and since the integrand is alternately positive and negative on the intervals $(0, \pi/2(n+1))$, $(\pi/2(n+1), 2\pi/2(n+1))$, etc., we see that S_n has (relative to $[0, \pi/2]$) alternately a minimum at 0, a maximum at $\pi/2(n+1)$, a minimum at $2\pi/2(n+1)$, and so on. Moreover, since $\sin 2(n+1)t$ has oscillations which are equal in magnitude, and is then multiplied by $1/\sin t$ which is strictly decreasing on $[0, \frac{1}{2}\pi]$, it follows that the largest among all the maxima of S_n is when $x = \pi/2(n+1)$, followed by the smallest of the minima (excluding 0) which is at $x = 2\pi/2(n+1)$, followed by the next largest maximum at $x = 3\pi/2(n+1)$, and so on (see Fig. 2.3). Hence on $(0, \pi/2]$ we have

$$0 < S_n(x) \leqslant S_n(\pi/2(n+1)) = \sum_{r=0}^{n} \frac{1}{2r+1} \sin\left(\frac{(2r+1)\pi}{2(n+1)}\right) = \alpha_n \text{ say.}$$

To prove our result, we have to show

(i) that $\alpha_n < \frac{1}{2}\pi$ for all n, and

(ii) that $\alpha_n \to \frac{1}{2} \int_0^\pi \frac{\sin x}{x} dx$, as $n \to \infty$.

To prove this, we make use of the definition of the integral (Definition A.9′). We put $x_0 = 0$, $x_1 = \pi/(n+1), \ldots, x_r = r\pi/(n+1), \ldots, x_{n+1} = \pi$; and

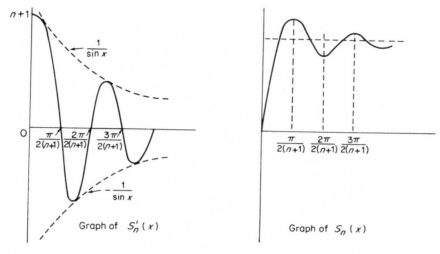

Fig. 2.3.

$y_r = \frac{1}{2}(x_r + x_{r+1}) = (2r+1)\pi/2(n+1)$: the points (x_r) form a dissection of $[0, \pi]$, and the points (y_r) lie in the intervals (x_r, x_{r+1}).

With this notation we may write

$$\alpha_n = \sum_{r=0}^{n} \frac{1}{2r+1} \sin \frac{(2r+1)\pi}{2(n+1)}$$

$$= \sum_{r=0}^{n} \frac{\pi}{2(n+1)} \frac{\sin(y_r)}{y_r} = \frac{1}{2} \sum_{r=0}^{n} (x_{r+1} - x_r) \frac{\sin(y_r)}{y_r}.$$

Since

$$\frac{\sin(y_r)}{y_r} < 1$$

for all y_r, we deduce at once that

$$\alpha_n < \frac{1}{2} \sum_{r=0}^{n} (x_{r+1} - x_r) = \frac{1}{2}\pi.$$

In addition, we see that the sum

$$\sum_{r=0}^{n} (x_{r+1} - x_r) \frac{\sin(y_r)}{y_r}$$

is a Reimann sum corresponding to the function $\sin\theta/\theta$ on the interval $[0, \pi]$, and that consequently it tends to $\int_0^\pi (\sin\theta/\theta)\,d\theta$ as $n \to \infty$: part (ii) follows at once from this

Notes (i) A more refined argument would show that in fact (α_n) is a *decreasing* sequence, but we do not need this result.

40

(ii) It is easy to extend the above argument to show that the first minimum value of S_n at $x = 2\pi/2(n+1)$ tends to $\frac{1}{2}\int_0^{2\pi}(\sin\theta/\theta)\,d\theta$, and that generally the kth stationary point (maximum or minimum, depending on the parity of k) at $x = k\pi/2(n+1)$ tends to $\frac{1}{2}\int_0^{k\pi}(\sin\theta/\theta)\,d\theta$. These numbers are alternately greater and less than the value $\frac{1}{2}\int_0^{\infty}(\sin\theta/\theta)\,d\theta = \frac{1}{4}\pi$ to which $S_n(x)$ converges as $n \to \infty$ for all fixed x in $(0, \pi)$. We have already seen that

$$\frac{4}{\pi}\sum_{r=0}^{\infty}\frac{1}{2r+1}\sin(2r+1)x$$

is the Fourier series of the function which is $+1$ on $(0, \pi)$, -1 on $(-\pi, 0)$, and equal to zero when x is a multiple of π. Theorem 2.3 shows that the series converges to f at every point, and uniformly on any interval of the form $[\delta, \pi - \delta]$. The behaviour of the first maximum of S_n (or the first minimum, etc.) as described above is not inconsistent with the uniformity of convergence, since for a given $\delta > 0$, and a fixed integer k, the position of the kth stationary point of S_n is at $k\pi/2(n+1)$, and for large enough n this will lie outside $[\delta, \pi - \delta]$.

We showed in Note (ii) above that as n increases, the values at successive maxima and minima approach limiting values which are alternately above and below the value of f to which S_n converges. In particular, the ratio of the limiting value at the first maximum to the value of f is

$$\frac{4}{\pi}\frac{1}{2}\int_0^{\pi}\frac{\sin\theta}{\theta}\,d\theta = 1.17898\cdots,$$

so that there is an 'overshoot' of approximately 18% in the limit. Similarly, the value at the first minimum approaches $0.90282\cdots$, there is an 'undershoot' of about 10%: this is indicated in Fig. 2.4.

This behaviour is known as the *Gibbs phenomenon*. It is an interesting sidelight on the development of the subject of Fourier series that the phenomenon was first discovered experimentally, by an apparatus (a harmonic analyser) which

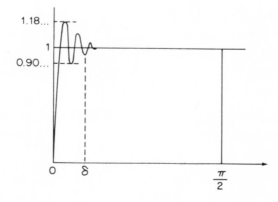

Fig. 2.4.

decomposed a given function into its sine and cosine components, and when first noticed was attributed to experimental error. For further details one can read the account in Lanczos (1966).

We have analysed the Gibbs phenomenon for a particular function for which the partial sums can be dealt with explicitly. However, it is easy to deduce that the same behaviour occurs in the neighbourhood of any simple discontinuity of the type described in the following result.

Theorem 2.9 Let f be a real-valued 2π-periodic FC-function, and suppose that for some x_0 in $[0, 2\pi]$ and $h > 0$, f is Lipschitz continuous on $(x_0 - h, x_0)$ and on $(x_0, x_0 + h)$, and has unequal left- and right-hand limits l_-, l_+ respectively at x. Let $\Delta = |l_+ - l_-|$.

Then the partial sums of the Fourier series for f display the Gibbs phenomenon in the neighbourhood of x_0: more exactly, as $n \to \infty$, the difference between the largest and smallest values which the partial sums of the Fourier series of f attain on a neighbourhood of x_0, approaches

$$\Delta \frac{2}{\pi} \int_0^\pi \frac{\sin \theta}{\theta} \, d\theta.$$

(The existence of the limits l_+ and l_- is in fact a direct consequence of the Lipschitz continuity on $(x_0, x_0 + h), (x_0 - h, x_0)$ respectively: we shall not stop to prove this since in examples the values of the limits are always evident.)

Proof Let ϕ be the function equal to 1, on $(x_0, x_0 + \pi)$, to -1 on $(x_0 - \pi, x_0)$, and 0 elsewhere: evidently ϕ is a translate of the function investigated in Lemma 2.8.

For $x \neq x_0$, let $g(x) = f(x) - \frac{1}{2}(l_+ - l_-)\phi(x)$. Then g has right- and left-hand limits at x_0, equal to $l_+ - \frac{1}{2}(l_+ - l_-)$ and $l_- + \frac{1}{2}(l_+ - l_-)$ respectively, both of which equal $\frac{1}{2}(l_+ + l_-)$.

Hence if we define $g(x_0) = \frac{1}{2}(l_+ + l_-)$ then g is an FC-function which by our hypotheses is Lipschitz continuous on $(x_0 - h, x_0 + h)$. Hence by Theorem 2.5, $S_n(g) \to g$ uniformly on any closed subinterval of $(x_0 - h, x_0 + h)$. Since $f = g + c\phi$ (writing $c = \frac{1}{2}(l_+ - l_-)$), we can write $S_n(f) = g(x_0) + (g(x) - g(x_0)) + (S_n(g)(x) - g(x)) + cS_n(\phi)$. The term $g(x_0)$ is a constant and since g is continuous at x_0, we can choose an interval about x_0 on which $g(x) - g(x_0)$ is as small as required. We have just shown that on such an interval $(S_n(g) - g)$ is uniformly small, and hence the behaviour of $S_n(f)$ is determined by that of $cS_n(\phi)$, i.e. the sums $S_n(f)$ display the Gibbs phenomenon near x_0, and the oscillation is c times the oscillation of ϕ, which is

$$\tfrac{1}{2}|l_+ - l_-| \text{ multiplied by } 2\frac{2}{\pi} \int_0^\pi \frac{\sin \theta}{\theta} \, d\theta, \text{ as required.}$$

2.4 PIECEWISE MONOTONE FUNCTIONS

Many functions which occur in applications may be continuous, in the usual sense of Definition A.1, but not Lipschitz continuous. We shall see in the next section that continuity alone does not ensure the convergence of the Fourier series. However, an alternative hypothesis, namely monotonicity, can often be used to give results about convergence and has the advantage that it can be easily recognised in examples. For instance, $f(x) = \sqrt{|x|}$ on $[-\pi, \pi]$, and $f(x) = 1/\log(1/|x|)$ on $[-\pi, \pi]$ $(f(0) = 0)$ are continuous, but not Lipschitz continuous near $x = 0$: both are monotone on $[-\pi, 0]$, and $[0, \pi]$, however. On the other hand, $f(x) = x^2 \sin(\pi^2/x)$ $(f(0) = 0$ again) shows that a function can be Lipschitz continuous (this one has a bounded derivative on $[-\pi, \pi]$) without being monotone, or even piecewise monotone (see Definition 2.12 below for the exact definition). More elaborate examples show that a Lipschitz continuous function need not be monotone on any interval. Thus, while many functions may be considered by either the methods of this section, or of Section 2.2, neither theory is wholly contained within the other.

More advanced readers will know that a function which is Lipschitz continuous on an interval must be of bounded variation and hence may be expressed as a difference of monotone functions there. However, we shall make essential use of the localization principle (Corollary 2.7) to prove the results of the present section, so that even with this extra knowledge the results of Section 2.2 are not subsumed in the present section.

We begin with a couple of preliminary results, the first of which is simply a standard result from integration theory (the 'second mean value theorem for integrals'). Since we shall only use it in this section it seems logical to include it here, rather than in the appendix.

Lemma 2.10 (i) Let f and g be real-valued FC-functions on $[a, b]$, and let f be monotone on $[a, b]$. Then for some c, $a \leqslant c \leqslant b$,

$$\int_a^b f(x)g(x)\,dx = f(a)\int_a^c g(x)\,dx + f(b)\int_a^b g(x)\,dx.$$

(ii) Let f and g be as in (i), and suppose that in addition f is positive and increasing on $[a, b]$. Then for some c, $a \leqslant c \leqslant b$,

$$\int_a^b f(x)g(x)\,dx = f(b)\int_c^b g(x)\,dx.$$

(iii) Let f be as in (ii) while g is a real- or complex-valued FC-function on $[a, b]$. Let $G = \sup\{|\int_c^b g(x)\,dx|; a \leqslant c \leqslant b\}$. Then

$$\left|\int_a^b f(x)g(x)\,dx\right| \leqslant f(b)G.$$

Proof We shall in fact need only (ii), so we prove this, leaving the deduction of (i) and (iii) as exercises (see exercise 6 at the end of this chapter).

Let $G(x) = \int_x^b g(t)\,dt$: G is continuous on $[a, b]$, so we can let G_1, G_2 be respectively the greatest and least values attained by G on the interval, and the result to be proved is equivalent to the statement that $\int_a^b f(x)g(x)\,dx$ lies between $f(b)G_1$ and $f(b)G_2$.

Let $M = \sup\{|g(x)|: a \leqslant x \leqslant b\}$, and let $D = \{x_0, x_1, \ldots, x_n\}$ be any dissection of $[a, b]$ (see Definition A.8). Let $\alpha_0 = f(x_0)$, and $\alpha_j = f(x_j) - f(x_{j-1})$ for $j = 1, 2, 3, \ldots, n$. Each α_j is non-negative, and $\sum_{j=0}^n \alpha_j = f(b)$.

Define H (Heaviside's function) by

$$H(x) = \begin{cases} 1 & \text{if } x \geqslant 0 \\ 0 & \text{if } x < 0 \end{cases}$$

and let $\phi(x) = \sum_{j=0}^n \alpha_j H(x - x_j)$ (see Fig. 2.5 for the graph of ϕ). Then

$$\left| \int_a^b f(x)g(x)\,dx - \int_a^b \phi(x)g(x)\,dx \right| = \left| \sum_{j=1}^n \int_{x_{j-1}}^{x_j} (f(x) - \phi(x))g(x)\,dx \right|$$

$$\leqslant \sum_{j=1}^n \int_{x_{j-1}}^{x_j} |f(x) - \phi(x)||g(x)|\,dx$$

$$\leqslant M \sum_{j=1}^n \alpha_j(x_j - x_{j-1})$$

$$\leqslant M f(b)\|D\|,$$

where as usual $\|D\| = \max_j (x_j - x_{j-1})$.

Hence $\int_a^b f(x)g(x)\,dx$ can be approximated arbitrarily closely by $\int_a^b \phi(x)g(x)\,dx$ by making $\|D\|$ small, and it is sufficient to show that $\int_a^b \phi(x)g(x)\,dx$ lies between $f(b)G_1$ and $f(b)G_2$. However,

$$\int_a^b \phi(x)g(x)\,dx = \sum_{j=0}^n \alpha_j \int_a^b H(x - x_j)g(x)\,dx$$

$$= \sum_{j=0}^n \alpha_j \int_{x_j}^b g(x)\,dx = \sum_{j=0}^n \alpha_j G(x_j).$$

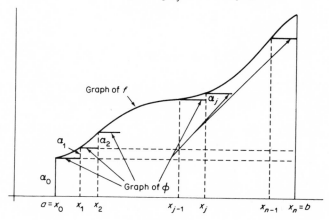

Fig. 2.5.

44

Hence

$$\frac{1}{f(b)} \sum_{j=0}^{n} \alpha_j G(x_j)$$

is a weighted average of values of the function G (that is, a linear combination whose coefficients are positive and sum to 1), and so must lie between the greatest and least values of G, as required.

Lemma 2.11   Let D_n be the Dirichlet kernel (Definition 2.2). Then there exists an *absolute* constant C ($4(2\pi + 1)$ will do) such that for any x, y in $[0, 2\pi]$.

$$\left| \int_x^y D_n(t)\,dt \right| \leqslant C.$$

Proof   Recall from Lemma 2.1 and Definition 2.2 that

$$D_n(t) = \frac{\sin(n + \frac{1}{2})t}{\sin\frac{1}{2}t} \qquad \text{and that } |D_n(t)| \leqslant 2n + 1 \text{ for all } t.$$

If we put $\delta_n = \pi/(n + \frac{1}{2})$ (the first zero of D_n), and for $0 < x \leqslant \pi$, write

$$\int_0^x D_n(t)\,dt = \int_0^{\delta_n} D_n(t)\,dt + \int_{\delta_n}^x D_n(t)\,dt,$$

then we see at once that

$$0 < \int_0^{\delta_n} D_n(t)\,dt < \delta_n(2n + 1) = 2\pi.$$

(If $x \leqslant \delta_n$, this estimate applies to $\int_0^x$ also.)

Since $(\sin\frac{1}{2}t)^{-1}$ is decreasing on $[\delta_n, x]$ we can apply the second mean value theorem (Lemma 2.10(ii)) to obtain

$$\int_{\delta_n}^x D_n(t)\,dt = \int_{\delta_n}^x \sin(n + \tfrac{1}{2})t/\sin\tfrac{1}{2}t\,dt$$

$$= (\sin\tfrac{1}{2}\delta_n)^{-1} \int_{\delta_n}^y \sin(n + \tfrac{1}{2})t\,dt \qquad \text{for some } y \text{ in } [\delta_n, x],$$

$$= (\sin\tfrac{1}{2}\delta_n)^{-1} \frac{1}{(n + \frac{1}{2})} [\cos(n + \tfrac{1}{2}t)]_{\delta_n}^y,$$

and this expression is in modulus at most

$$(\sin\tfrac{1}{2}\delta_n)^{-1} \frac{2}{(n + \frac{1}{2})} = 2\{(n + \tfrac{1}{2})\sin(\pi/2n + 1)\}^{-1}.$$

However, the elementary inequality

$$\sin\theta > \frac{2}{\pi}\theta \text{ for } 0 < \theta < \frac{\pi}{2}$$

shows that

$$(n + \tfrac{1}{2}) \sin\left(\frac{\pi}{2n+1}\right) > (n + \tfrac{1}{2})\frac{2}{2n+1} = 1.$$

Combining these results we have shown that if $0 < x \leqslant \pi$, then $|\int_0^x D_n(t)\,dt| \leqslant 2\pi + 2$, and the Lemma follows from this and the facts that $\int_x^{2\pi} D_n = \int_0^{2\pi} D_n - \int_0^{2\pi-x} D_n = 2\pi - \int_0^{2\pi-x} D_n$, and

$$\int_x^y D_n = \int_0^y D_n - \int_0^x D_n.$$

Note An alternative argument based on the geometrical properties of D_n (cf. the proof of Lemma 2.8) shows that if $0 < x \leqslant 2\pi$, then $0 < \int_0^x D_n \leqslant 2\pi$, and hence that $|\int_y^x D_n| \leqslant 2\pi$.

Having established these lemmas we can define the class of functions with which we shall be concerned, and prove our main convergence theorem for them.

Definition 2.12   Let f be a real-valued function defined on an interval $[a, b]$. We say f is *piecewise monotone* on $[a, b]$ if $[a, b]$ can be written as the union of a finite number of intervals, $[x_{i-1}, x_i]$, $i = 1, 2, \ldots, n$ say, such that f is monotone on each open interval (x_{i-1}, x_1). (Notice that the values of f at the points x_i are not relevant to the definition. The function may of course be monotone increasing on some subintervals, and monotone decreasing on others.)

The examples 2.4(i), (ii), and (iii) are piecewise monotone, while 2.4(iv) is not. For piecewise monotone functions a situation analogous to that described in Theorem 2.3 and 2.5 occurs: the Fourier series converges to f at points of continuity (with uniformity on closed proper subintervals of continuity) while at a jump discontinuity (the only kind possible in this context) it converges to the average of the left- and right-hand limits.

By restricting f to each of the subintervals (x_{i-1}, x_i) in turn, we see that a piecewise monotone function can be written as a finite sum of functions which are monotone on a single interval, and zero elsewhere. Our overall restriction to FC-functions means that by a further sub-division if necessary we can assume that f is continuous on each (x_{i-1}, x_i). Our convergence result (Theorem 2.14 below) will thus follow from the following theorem.

Theorem 2.13   Let (a, b) be a subinterval of $[0, 2\pi]$ and let f be a bounded real-valued function which is continuous and monotone on (a, b), and zero elsewhere on $[0, 2\pi]$.

Then $S(f)$ converges to f at each point of (a, b), the convergence being uniform on subintervals of the form $[a + \delta, b - \delta]$, $\delta > 0$. At a, b the series converges to $\tfrac{1}{2} l_1, \tfrac{1}{2} l_2$ respectively, where

$$l_1 = \lim_{x \to a_+} f(x), \quad l_2 = \lim_{x \to b_-} f(x).$$

Note The convergence of $S(f)$ to zero at points disjoint from $[a, b]$ is assured by the results of Section 2.2.

Proof By considering $\pm f$ we may suppose that f is increasing on (a, b). Let x be a point of (a, b) and given $\varepsilon > 0$, choose $\delta > 0$ so that

$$|f(y) - f(x)| \leqslant \varepsilon \text{ if } |y - x| \leqslant \delta.$$

Since $D_n(t) = D_n(-t)$ we can put $-t$ for t in Lemma 2.1 to obtain

$$S_n(f, x) = \frac{1}{2\pi} \int_0^{2\pi} f(x + t) D_n(t) \, dt,$$

and hence

$$S_n(f, x) - f(x) = \frac{1}{2\pi} \int_0^{2\pi} (f(x + t) - f(x)) D_n(t) \, dt.$$

We now divide the range of integration into $[0, \delta]$, $[\delta, 2\pi - \delta]$, and $[2\pi - \delta, 2\pi]$. The hypotheses of f ensure that $f(x + t) - f(x)$ is positive and increasing (as a function of t) on $[0, \delta]$, and hence that by Lemma 2.10 (iii)

$$\int_0^\delta (f(x + t) - f(x)) D_n(t) \, dt = (f(x + \delta) - f(x)) \int_{\delta_1}^\delta D_n(t) \, dt$$

for some δ_1 with $0 < \delta_1 < \delta$.

But $|f(x + \delta) - f(x)| \leqslant \varepsilon$, and $|\int_{\delta_1}^\delta D_n(t) \, dt| \leqslant C$ by Lemma 2.11, so that

$$\left| \frac{1}{2\pi} \int_0^\delta (f(x + t) - f(x)) D_n(t) \, dt \right| \leqslant \frac{C}{2\pi} \varepsilon.$$

A similar argument shows that

$$\left| \frac{1}{2\pi} \int_{2\pi - \delta}^{2\pi} (f(x + t) - f(x)) D_n(t) \, dt \right| \leqslant \frac{C}{2\pi} \varepsilon.$$

In addition we have

$$\frac{1}{2\pi} \int_\delta^{2\pi - \delta} (f(x + t) - f(x)) D_n(t) \, dt$$

$$= \frac{1}{2\pi} \int_0^{2\pi} \phi(t) \sin(n + \tfrac{1}{2}) t \, dt$$

where $\phi(t) = (f(x + t) - f(x))/\sin \tfrac{1}{2} t$ on $[\delta, 2\pi - \delta]$, and is zero elsewhere.

Then ϕ is an FC-function and thus the integral of $\phi(t) \sin(n + \tfrac{1}{2}) t$ tends to zero as $n \to \infty$ by the Riemann–Lebesgue lemma. This proves that $S_n(f, x) \to f(x)$ as $n \to \infty$. To obtain the uniform convergence, notice that f will be uniformly continuous on a closed subinterval of (a, b), and hence that the estimates obtained above for the integrals $\int_0^\delta, \int_{2\pi - \delta}^{2\pi}$ may be made uniform in x. The integral

$\int_0^{2\pi}\phi(t)\sin(n+\frac{1}{2})t\,dt$ may then be shown to tend uniformly to zero in the same way as was done in the proof of Theorem 2.5: we shall not repeat the details.

To show convergence to $\frac{1}{2}l_1$ at a we write

$$S_n(f,a) = \frac{1}{2\pi}\int_0^{2\pi} f(a-t)D_n(t)\,dt$$

$$= \frac{1}{2\pi}\int_0^{2\pi} \tfrac{1}{2}(f(a-t)+f(a+t)D_n(t)\,dt,$$

and thus

$$S_n(f,a) - \tfrac{1}{2}l_1 = \frac{1}{2\pi}\int_0^{2\pi} \tfrac{1}{2}(f(a+t)+f(a-t)-l_1)D_n(t)\,dt.$$

Given $\varepsilon > 0$, choose $\delta > 0$ so that $|f(a+t)-l_1| < \varepsilon$ and $f(a-t) = 0$ if $0 < t < \delta$. Then

$$\left| \frac{1}{2\pi}\int_0^{\delta} \tfrac{1}{2}(f(a-t)+f(a+t)-l_1)D_n(t)\,dt \right|$$

$$= \left| \frac{1}{2\pi}\int_0^{\delta} \tfrac{1}{2}(f(a+t)-l_1)D_n(t)\,dt \right|$$

$$= \frac{1}{2\pi}\frac{1}{2}\left|f(a+\delta)-l_1\right|\left|\int_{\delta_1}^{\delta} D_n(t)\,dt\right|$$

$$< \frac{1}{2\pi}\frac{1}{2}\varepsilon C = \frac{1}{4\pi}\varepsilon C.$$

In the same way we show that the integral over $[2\pi - \delta, 2\pi]$ is less than $C\varepsilon/4\pi$ and the integral over $[\delta, 2\pi - \delta]$ tends to zero as $n \to \infty$ as before. The argument showing convergence to $\frac{1}{2}l_2$ at b is similar, and this completes the proof of Theorem 2.13.

Our principal result on piecewise monotone functions now follows at once.

Theorem 2.14 Let f be a 2π-periodic FC-function which is piecewise monotone on some subinterval $[a, b]$ of $[0, 2\pi]$. Then the Fourier series of f is convergent at each point x of (a, b) and its sum is $\frac{1}{2}\lim_{h\to 0}(f(x+h)+f(x-h))$: it is uniformly convergent to f on each closed subinterval of (a, b) on which f is continuous.

Proof The hypotheses on f imply that it may be written in the form

$$f = f_0 + \sum_{j=1}^{n} f_j, \quad \text{where } f_0 = \begin{cases} 0 & \text{on } [a, b] \\ f & \text{elsewhere} \end{cases}$$

while each f_j $(1 \leqslant j \leqslant n)$ is monotone on some subinterval of $[a, b]$, and zero elsewhere. The localization principle (Corollary 2.7) shows that each of these functions has a Fourier series which converges uniformly with sum zero, on the intervals on which they vanish. The result follows from this, and from Theorem 2.13.

Examples 2.15 (i) Many of our earlier examples fall within the scope of Theorem 2.14: for instance, those in Example 2.4(i), (ii), and (iii).

(ii)   Let $f(x) = \sqrt{|x|}$ on $[-\pi, \pi]$.

The Fourier coefficients cannot be explicitly computed here, and the function is not Lipschitz continuous at $x = 0$, so that Theorem 2.3 does not apply. However, f is continuous and piecewise monotone on $[-\pi, \pi]$, so Theorem 2.14 ensures that the Fourier series converges uniformly with sum f. Similar observations apply to any continuous even function of this type, for example.

$$f(x) = (\log(2\pi/|x|))^{-1} \quad \text{if} \quad 0 < |x| \leqslant \pi, \quad f(0) = 0.$$

Note   The crucial step in the proof of Theorem 2.13 was the statement that if f was positive and increasing on $[0, \delta]$, then

$$\left| \int_0^\delta f(t) D_n(t) \, dt \right| \leqslant f(\delta) C.$$

The reader may well ask why it was necessary to use complicated arguments such as that in Lemma 2.11, when one could instead estimate the above integral by the obvious bound $f(\delta) \int_0^\delta |D_n(t)| \, dt$.

This estimate is insufficient because in fact the integral of $|D_n|$ over any such interval tends to infinity with n, and we briefly indicate why this is so.

Consider for example $\int_0^\pi |D_n(t)| \, dt = \int_0^\pi |\sin(n + \frac{1}{2})t / \sin \frac{1}{2}t| \, dt$. On each interval of the form

$$\left(\frac{(j-1)\pi}{n + \frac{1}{2}}, \frac{j\pi}{n + \frac{1}{2}} \right),$$

$\sin \frac{1}{2}t$ is decreasing, while $\sin(n + \frac{1}{2})t$ has constant sign (Fig. 2.1). Hence the integral of $|D_n|$ over this interval is at least

$$\left(\sin\left(\frac{j\pi}{2n+1} \right) \right)^{-1} \int_{(j-1)\pi/(n+\frac{1}{2})}^{j\pi/(n+\frac{1}{2})} |\sin(n + \frac{1}{2})t| \, dt$$

$$= \frac{2}{n + \frac{1}{2}} \left(\sin\left(\frac{j\pi}{2n+1} \right) \right)^{-1}, \qquad \text{for } j = 1, 2, \dots, n.$$

But since $\sin \theta < \theta$ for $\theta > 0$, this is greater than

$$\frac{2}{n + \frac{1}{2}} \frac{2n+1}{j\pi} = \frac{4}{\pi j}.$$

Hence

$$\int_0^\pi |D_n(t)| / \, dt > \frac{4}{\pi} \sum_{j=1}^n \frac{1}{j}, \quad \text{which tends to infinity with } n.$$

2.5 DIVERGENCE OF FOURIER SERIES

For more than half a century after Fourier's first systematic study of trigonometric series, the possibility remained open that the Fourier series of any continuous function f might be convergent everywhere with sum f.

The first example of a continuous function whose Fourier series was divergent at a point was found by du Bois-Reymond in 1873. We shall give a simplified version of his construction in Theorem 2.16. A further possibility was that there might be continuous functions whose Fourier series was divergent at *every* point – indeed, A. Kolmogorov showed in the mid-1930s that there was a function which was integrable in the sense of Lebesgue whose Fourier series was divergent everywhere. It was not until 1966 that L. Carleson showed that for any f whose square was integrable (in particular, for any continuous function) its Fourier series must be convergent 'almost everywhere' – that is except on a set which has Lebesgue measure zero. The precise relationship between the continuity or integrability of a function and the sets on which its Fourier series may converge is still not fully known, and remains one of the main problems of contemporary harmonic analysis. Some references to the advanced literature in this area can be found in the Guide to Further Reading in Appendix C: for the present we shall content ourselves with the following example.

Theorem 2.16 There exists a continuous function whose Fourier series is divergent at the origin.

Proof Let $S_n(x) = \sum_{j=1}^{n} (1/j) \sin jx$, for real x and $n = 1, 2, 3, \ldots$.
We may differentiate to obtain

$$S_n'(x) = \sum_{j=1}^{n} \cos jx = \tfrac{1}{2}(D_n(x) - 1), \text{ and hence}$$

$$S_n(x) = \frac{1}{2}\int_0^x D_n(t)\,dt - \tfrac{1}{2}x, \text{ since } S_n(0) = 0.$$

The proof of Lemma 2.11 shows that if $0 \leqslant x \leqslant 2\pi$, we have

$$\left| \int_0^x D_n(t)\,dt \right| < C,$$

and hence $|S_n(x)| < \tfrac{1}{2}(\pi + C)$ for $0 \leqslant x \leqslant 2\pi$,
i.e. (S_n) is a uniformly bounded sequence of functions on $[0, 2\pi]$.
Now, for N, n integers, with $0 < n < N$, consider

$$F(N, n, x) = 2 \sin Nx\, S_n(x)$$

$$= 2 \sin Nx \sum_{j=1}^{n} \frac{1}{j} \sin jx$$

$$= \sum_{j=1}^{n} \frac{1}{j} (\cos (N - j)x - \cos (N + j)x).$$

This is a trigonometric polynomial, here a sum of cosines whose index lies between $N - n$ and $N + n$. Our estimate for S_n shows that $F(N, n, x)$ is uniformly bounded for all N, n, and x.

We now choose increasing sequences $(N_1, N_2, \ldots, N_r, \ldots)$ and $(n_1, n_2, \ldots, n_r, \ldots)$ to be specified presently: for the moment we require only that for each r,

$$N_r + n_r < N_{r+1} - n_{r+1},$$

so that the indices of the terms in $F(N_r, n_r, x)$ for different values of r do not overlap.

Now consider the series

$$\sum_{r=1}^{\infty} \frac{1}{r^2} F(N_r, n_r, x):$$

the continuity and uniform boundedness of the F's shows that it is uniformly convergent (by the M-test, Theorem A.23) and hence that its sum, which we denote by $f(x)$, is continuous. In fact $f(0) = 0$ since $S_n(0) = 0$. The uniform convergence also shows that the series obtained from the above by writing each $F(N_r, n_r, x)$ as a sum of cosines and putting the whole in order of increasing index is in fact just the Fourier series of f.

If we put $x = 0$, the sum of the Fourier series up to terms of order $N_r + n_r$ is 0 since $F(N_r, n_r, 0) = 0$: however, the sum up to terms of order N_r differs from this by

$$\frac{1}{r^2} \sum_{j=1}^{n_r} \frac{1}{j} > \frac{1}{r^2} \log(n_r).$$

Hence if we choose $n_r = 2^{r^2}$ then $(1/r^2) \log(n_r) = \log 2$ does not tend to zero, so that the series does not satisfy the Cauchy criterion (stated for sequences in Theorem A.20) and thus cannot converge.

It remains to show that with this choice of n_r we can choose N_r to satisfy the no-overlapping condition. We see that $N_r = 2n_r = 2^{r^2+1}$ will do, since

$$(N_{r+1} - n_{r+1}) - (N_r + n_r)$$
$$= 2^{(r+1)^2+1} - 2^{(r+1)^2} - 2^{r^2+1} - 2^{r^2}$$
$$= 2^{(r+1)^2} - 3 \times 2^{r^2} = 2^{r^2}(2^{2r+1} - 3) > 0$$

(since $2^{2r+1} \geqslant 2^3 = 8 > 3$). This completes the required construction.

Note The distinction, which is vital to the above construction, between the series which defines f, namely $\sum_{r=1}^{\infty} (1/r^2) F(N_r, n_r, x)$, and the Fourier series of f is often confusing on first acquaintance. To reverse the procedure followed above we may say that the former is obtained from the latter by grouping terms together, or equivalently by inserting brackets, and it is well known in elementary analysis that this may turn a divergent series into a convergent one: a trivial example is the divergent series

$$1 - 1 + 1 - 1 + 1 - 1 + \cdots$$

which becomes convergent on bracketing consecutive pairs of terms together. The above construction is of course more subtle: the reader is encouraged to work out the first few values of n_r, N_r and the first terms of the Fourier series of f explicitly (exercise 8 at the end of this chapter).

The above result can be extended to give an example of a continuous function whose Fourier series diverges at every rational point – for the details the reader should consult Rogosinski (1950, page 77). However, this is by no means the end of the matter, as the Guide to Further Reading indicates.

EXERCISES

1 For each of the following functions, use either Theorem 2.3 or Theorem 2.14 to find the points at which its Fourier series converges, and its sum at these points. (It is not necessary – or indeed possible – to calculate the Fourier coefficients explicitly in all cases.)

   (i) $f(x) = x$ for $-\pi < x < \pi$, $f(\pm \pi) = 0$;

   (ii) $f(x) = \begin{cases} x & \text{for } 0 < x < \pi, \\ 0 & \text{elsewhere}; \end{cases}$

   (iii) $f(x) = x^2$ for $-\pi \leqslant x \leqslant \pi$;

   (iv) $f(x) = \sqrt{\pi^2 - x^2}$ for $-\pi \leqslant x \leqslant \pi$;

   (v) $f(x) = x(\pi - x)$ for $0 \leqslant x \leqslant \pi$, and f is extended to $[-\pi, \pi]$ (a) as an even function, (b) as an odd function;

   (vi) $f(x) = x \sin(\pi^2/x)$ for $0 < |x| \leqslant \pi$, $f(0) = 0$.

2 For each of parts (i)–(v) of exercise 1, find the intervals on which the Fourier series in uniformly convergent. (The question of the uniform convergence in part (vi) in the neighbourhood of $x = 0$ cannot be decided on the basis of our results.)

3 Let f be any 2π-periodic FC-function, with Fourier series

$$S(f) = \sum_{n=-\infty}^{\infty} c_n e^{inx}$$

$$= \tfrac{1}{2}a_0 + \sum_{n=1}^{\infty} (a_n \cos nx + b_n \sin nx).$$

Let $F(x) = \int_0^x f(t)\,dt - c_0 x$: show that F is continuous and 2π-periodic and that

$$S(F) = C_0 + \sum_{\substack{n=-\infty \\ n \neq 0}}^{\infty} \frac{c_n}{in} e^{inx}$$

$$= \tfrac{1}{2}A_0 + \sum_{n=1}^{\infty} \frac{1}{n}(a_n \sin nx - b_n \cos nx),$$

$$\text{where } \tfrac{1}{2}A_0 = C_0 = \frac{1}{2\pi}\int_0^{2\pi} x(c_0 - f(x))\,dx,$$

(*Hint*: integrate by parts to find the Fourier coefficients of F.)
Thus $S(F)$ can be obtained from $S(f)$ by formally integrating termwise.

Use Theorem 2.3 to show that $S(F)$ is uniformly convergent on $[0, 2\pi]$ with sum F.

4 Deduce from exercise 3 above that for any 2π-periodic FC-function f, the series $\sum_{n=1}^{\infty}(1/n)b_n$ is convergent, with sum C_0.

5 (i) Apply exercise 4 to the function of Example 2.4 (ii) to find $\sum_{n=1}^{\infty}(1/n^2)$. (This gives an alternative derivation to the one used in Chapter 1.)

 (ii) Use exercise 4, and elementary fact that $\sum_{n=1}^{\infty} 1/n \log(n+1)$ is divergent (by the integral test) to show that the series $\sum_{n=1}^{\infty} \sin nx/\log(n+1)$ is not the Fourier series of any FC-function. (We shall see in exercise 11 below that this series is in fact convergent for all x – however, its sum is not an FC-function, (or even integrable, though a proof of this is outside our scope.)

6. If f is monotone on $[a, b]$, then $\pm(f(x) - f(a))$ is positive and increasing on $[a, b]$. Use this fact to deduce Lemma 2.10 (i) from 2.10 (ii).

To deduce Lemma 2.10 (iii) use the expression

$$\int_a^b \phi(x)g(x)\,dx = \sum_{j=0}^{n} \alpha_j G(x_j) \text{ obtained in the proof of 2.10(ii):}$$

this is increased in modulus if each $G(x_j)$ is replaced by G.

7 Let f be a piecewise monotone function on $[0, 2\pi]$: show that for some constant K, $|C_n| \leqslant K/|n|$ for all $n \pm 0$.

Hint: Apply Lemma 2.10(i) with $g(x) = e^{inx}$ on each interval on which f is monotone.

8 In the construction used in Theorem 2.16, calculate $n_1, n_2, n_3, N_1, N_2, N_3$, and hence indicate the terms in the Fourier series, as far as the term in $\cos 1536x$.

9 Let $(a_n)_{n=0}^{\infty}, (b_n)_{n=0}^{\infty}$ be sequences of real or complex numbers, and let $A_0 = a_0$, $A_1 = a_0 + a_1$, and generally $A_n = a_0 + a_1 + \cdots + a_n$. Show by putting $a_m = A_m - A_{m-1}$ $(m \geqslant 1)$ that $\sum_{m=0}^{n} a_m b_m = \sum_{m=0}^{n-1} A_m(b_m - b_{m+1}) + A_n b_n$. Deduce that if (A_n) is bounded, and (b_n) decreases monotonically to zero, then the series $\sum_0^{\infty} a_n b_n$ is convergent.

10 Let
$$S_n(x) = 1 + \cos x + \cos 2x + \cdots + \cos nx,$$

$$T_n(x) = \sin x + \sin 2x + \cdots + \sin nx.$$

Show that for fixed x in the interval $(0, 2\pi)$, (S_n) and (T_n) are bounded as $n \to \infty$, and that the bound may be made uniform on any interval of the form $[\delta, 2\pi - \delta]$. (For S_n, this result follows from the expression for D_n in Definition 2.2(ii), and a similar calculation gives the result for T_n).

11 Combine the results of exercises 9 and 10 to show that if $(\alpha_n)_{n=0}^{\infty}$ is a sequence of positive numbers which decreases to zero as $n \to \infty$, then both $\sum_{n=0}^{\infty} \alpha_n \cos nx$ and $\sum_1^{\infty} \alpha_n \sin nx$ are convergent on $(0, 2\pi)$ and uniformly convergent on $[\delta, 2\pi - \delta]$. In particular, $\sum_{n=1}^{\infty} \sin nx/\log(n+1)$ is convergent on $[0, 2\pi]$, as was stated in exercise 5.

CHAPTER 3

The Dirichlet Problem and the Poisson Integral

3.1 HARMONIC FUNCTIONS

Among real-valued functions on $\mathbb{R}^2$ a particularly important role in both pure mathematics and its physical applications is played by the harmonic functions. These may be introduced as continuous functions with the *mean-value property*: that is, for any point $a = (a_1, a_2)$ and $r > 0$, we have

$$f(a_1, a_2) = \frac{1}{2\pi} \int_0^{2\pi} f(a_1 + r \cos \theta, a_2 + r \sin \theta) \, d\theta. \tag{3.1}$$

This says simply that for any point a and any circle centred at a, the value of f at a is equal to its average value on the circle. It is thus the two-dimensional analogue of the one-dimensional condition

$$f(x) = \tfrac{1}{2}\{f(x + h) + f(x - h)\}$$

which is easily shown to be equivalent to the linearity of f (see exercise 1 at the end of this chapter). In terms of derivatives this can be written $f''(x) = 0$: similarly, in two-dimensions it can be shown that condition (3.1) is equivalent (for functions with continuous partial derivatives) to Laplace's equation

$$\frac{\partial^2 f}{\partial x^2} + \frac{\partial^2 f}{\partial y^2} = 0. \tag{3.2}$$

An immediate proof of the equivalence of (3.1) and (3.2) would take us too far afield (for the implication $(3.2 \Rightarrow (3.1)$ see exercise 13 at the end of the chapter): instead we shall take equation (3.2) as our starting point.

The importance of equation (3.2) is due to its occurrence in many different branches of mathematics, and we shall outline several of them briefly. (No knowledge of these background areas in required in order to understand the rest of this chapter, however).

53

For an applied mathematician or natural scientist, equation (3.2) is simply the field equation which determines potential functions, whether they are gravitational, electrostatic, hydrodynamic, or of several other less common types. A potential function is a scalar function f, whose gradient is the vector function (denoted grad(f) or ∇f) defined (in two dimensions) by

$$\nabla f = \left(\frac{\partial f}{\partial x}, \frac{\partial f}{\partial y} \right).$$

This vector function (or its negative, according to the convection being used) determines a field of gravitational or electric force, or a direction of fluid flow, in the examples mentioned above. Equation (3.2) is then the condition that this field has zero divergence (the divergence of a vector function $F = (F_1, F_2)$ is $\partial F_1/\partial x + \partial F_2/\partial y$): in physical terms this occurs when there is no mass or electric charge in the region occupied by the field, or when the fluid is incompressible.

A pure mathematician may meet Laplace's equation for the first time as a property of the real and imaginary parts of an analytic function. For if $f(z) = u(x, y) + iv(x, y)$ has a complex derivative at each point on an open set G (here $z = x + iy$ is a point of G and u, v are real-valued functions), then it can be shown that u and v have continuous partial derivatives of all orders at all points of G, and that at every point the Cauchy–Riemann equations

$$\frac{\partial u}{\partial x} = \frac{\partial v}{\partial y}, \frac{\partial u}{\partial y} = -\frac{\partial v}{\partial x} \tag{3.3}$$

are satisfied. Equation (3.2) now follows at once for u (or v) by elimination:

$$\frac{\partial^2 u}{\partial x^2} = \frac{\partial^2 v}{\partial x \partial y} = \frac{\partial^2 v}{\partial y \partial x} = -\frac{\partial^2 u}{\partial y^2},$$

when the equality of the mixed partial derivatives is assured by their continuity. We shall use this as a method for constructing harmonic functions, and also show that it characterizes them: given any harmonic function there is always (locally at least) an analytic function of which it is the real part (Theorem 3.16 below).

Looked at from the point of view of the differential geometry of the surface in three-space given by the equation $z = f(x, y)$, Laplace's equation says simply that at each point the total curvature is zero (compare the analogous situation in one dimension where f must be a straight line). Thus at any point the bending of the surface in perpendicular directions is equal and opposite, a property of a surface which was first investigated by Gauss.

For all of these reasons the class of harmonic functions is of great importance: we shall see later in this chapter how our knowledge of Fourier series enables us to solve problems about harmonic functions which then lead back to a deeper knowledge of Fourier theory.

We begin our systematic study with the formal definition.

Definition 3.1 Let f be a real- or complex-valued function defined on an open set G in $\mathbb{R}^2$. (A set G is *open* if each of its points is surrounded by some disc which lies entirely inside G.) We shall say that f is *harmonic on G* if it has partial derivatives of first and second orders which are continuous and which satisfy, at all points of G, Laplace's equation

$$\frac{\partial^2 f}{\partial x^2} + \frac{\partial^2 f}{\partial y^2} = 0.$$

Examples 3.2   (i) For $n = 1, 2, 3, \ldots$, let

$$u(x, y) = \text{Re}(x + iy)^n, v(x, y) = \text{Im}(x + iy)^n.$$

Then u and v are harmonic throughout $\mathbb{R}^2$. For if $f(z) = z^n = u(x, y) + iv(x, y)$, then partial differentiation with respect to x and y shows that

$$nz^{n-1} = \frac{\partial u}{\partial x} + i\frac{\partial v}{\partial x}, \qquad niz^{n-1} = \frac{\partial u}{\partial y} + i\frac{\partial v}{\partial y}.$$

The Cauchy–Riemann equations (3.3), and thus Laplace's equation, now follow for u and v by equating real and imaginary parts.

(ii) The method indicated in (i) may be followed for any differentiable function f, as is indicated in Lemma 3.13 below. For example,

$f(z) = e^z$ gives $u(x, y) = e^x \cos y, v(x, y) = e^x \sin y$;

$f(z) = \sin z$ gives $u(x, y) = \sin x \cosh y, v(x, y) = \cos x \sinh y$;

and $f(z) = \log z$ (the principal value, defined then z is not real and negative to have imaginary part between $-\pi$ and π) gives $u(x, y) = \frac{1}{2}\log(x^2 + y^2)$ and $v(x, y)$ as the unique angle in $(-\pi, \pi)$ which satisfied $\cos v = x(x^2 + y^2)^{-1/2}$, $\sin v = y(x^2 + y^2)^{-1/2}$.

Note   In order to make the foregoing discussion and examples a little more systematic, we shall adopt the usual practice of identifying a point (x, y) in $\mathbb{R}^2$ with $x + iy$ in $\mathbb{C}$, and identifying similarly $f(x, y)$ and $f(x + iy)$. Thus for instance, equation (3.1) could more neatly be written as

$$f(a) = \frac{1}{2\pi}\int_0^{2\pi} f(a + re^{i\theta})\, d\theta,$$

while in Example 3.2 (ii) above,

$$u(x, y) = \text{Re}(\log z) = \log|z| \quad \text{and} \quad v(x, y) = \text{Im}(\log z) = \text{Arg}\, z.$$

If we write $z = x + iy = re^{i\theta}$, so that $z^n = r^n(\cos n\theta + i\sin n\theta)$, then Examples 3.2(i) and (ii) show that the functions given by $r^n \cos n\theta$ and $r^n \sin n\theta$ are harmonic throughout $\mathbb{R}^2$, while $\log r$ is harmonic except at the origin, and $\theta = \text{Arg}\, z$ is harmonic except on the negative real axis, where it is undefined.

We shall see in the next section that for circular regions at least, the supply of functions given by these examples and linear combinations of them will be sufficient to construct all harmonic functions.

3.2 THE DIRICHLET PROBLEM AND THE POISSON INTEGRAL

The simplest version of the Dirichlet problem which we shall consider is that of determining a harmonic function in a circular region whose values are prescribed on the boundary of the region. (In a physical problem, these values might be determined by a distribution of electrical charge around the boundary, for instance.) We suppose that we are given a function $f_1(z)$ whose values are known for all $z = e^{i\theta}$, $0 \leqslant \theta \leqslant 2\pi$, and our problem is to find a function $f(z)$ which is harmonic for $|z| < 1$, while $f(re^{i\theta}) \to f_1(e^{i\theta})$ when $r \to 1_-$.

In order to motivate our ensuing construction we suppose that the problem has a solution in the form of a *finite* sum of terms $r^n \cos n\theta$, $\sin n\theta$. That is, for some $N = 1, 2, 3, \ldots$, we can write

$$f(re^{i\theta}) = \tfrac{1}{2}A_0 + \sum_{n=1}^{N} r^n(A_n \cos n\theta + B_n \sin n\theta),$$

(where the factor $\tfrac{1}{2}$ in front of A_0 is in deference to the convention of Chapters 1 and 2). This is certainly a harmonic function, with

$$f_1(e^{i\theta}) = \tfrac{1}{2}A_0 + \sum_{n=1}^{N} (A_n \cos n\theta + B_n \sin n\theta).$$

We recognize at once that f_1 is a trigonometric polynomial and that A_n and B_n are its Fourier coefficients.

For a general function f_1 (not necessarily a trigonometric polynomial) we simply reverse our point of view. That is, we define the Fourier coefficients

$$a_n = \frac{1}{\pi} \int_0^{2\pi} f_1(e^{i\theta}) \cos n\theta \, d\theta, \qquad b_n = \frac{1}{\pi} \int_0^{2\pi} f_1(e^{i\theta}) \sin n\theta \, d\theta, n \geqslant 0,$$

as in Chapter 1, and hence *construct* the function

$$f(re^{i\theta}) = \tfrac{1}{2}a_0 + \sum_{n=1}^{\infty} r^n(a_n \cos n\theta + b_n \sin n\theta).$$

We shall show that the function f defined in this way is harmonic for $r < 1$ and has the required boundary behaviour, at least at points where f is continuous, thus obtaining a solution to the Dirichlet problem in the simplest case.

Definition 3.3 Let $f_1(e^{i\theta})$ be a real- or complex-valued FC-function defined for $0 \leqslant \theta \leqslant 2\pi$, and let $(a_n)_0^{\infty}, (b_n)_1^{\infty}$, be its Fourier coefficients. The function $f(re^{i\theta})$ which is defined by

$$f(re^{i\theta}) = \tfrac{1}{2}a_0 + \sum_1^\infty r^n(a_n \cos n\theta + b_n \sin n\theta)$$

is called the *Poisson integral* of f_1. We have

$$f(re^{i\theta}) = (f_1 * P_r)(e^{i\theta}),$$

where

$$P_r(e^{i\theta}) = \frac{1-r^2}{1-2r\cos\theta+r^2} = 1 + 2\sum_{n=1}^\infty r^n \cos n\theta \qquad (0 \leqslant r < 1)$$

is called the *Poisson kernel*. In terms of the complex coefficients

$$(c_n)\, f(re^{i\theta}) = c_0 + \sum_{n=1}^\infty r^n(c_n e^{inh} + c_{-n} e^{-in\theta}) = \sum_{-\infty}^\infty c_n r^{|n|} e^{in\theta}.$$

Note The assertions contained in this definition are proved in Lemma 3.4 below. The slight notational change between this chapter and the preceding ones – we write $f(e^{i\theta})$ where we previously wrote $f(\theta)$ for $0 \leqslant \theta \leqslant 2\pi$ – reflects the two-dimensional viewpoint of this chapter and should not cause any confusion; in any case we shall often use the notation interchangeably and write $P_r(\theta)$ for $P_r(e^{i\theta})$ when it seems appropriate.

Lemma 3.4 (i) For each r, $0 \leqslant r < 1$, the series $1 + 2\sum_{n=1}^\infty r^n \cos n\theta = P_r(\theta)$ is uniformly convergent on $[0, 2\pi]$, with sum

$$\frac{1-r^2}{1-2r\cos\theta+r^2}.$$

(ii) Let $f(re^{i\theta})$ be the Poisson integral of $f_1(r^{i\theta})$, as in Definition 3.3. Then $f(re^{i\theta}) = (f_1 * P_r)(\theta)$.

Proof (i) For fixed $r < 1$, the geometric series $\sum_1^\infty r^n$ is convergent, and hence $1 + 2\sum_1^\infty r^n \cos n\theta$ is uniformly convergent by the M-test. The sum can be found by writing it in the form

$$1 + \sum_1^\infty r^n(e^{in\theta} + e^{-in\theta}) = 1 + \sum_1^\infty \{(re^{i\theta})^n + (re^{-i\theta})^n\}$$

$$= 1 + re^{i\theta}(1 - re^{i\theta})^{-1} + re^{-i\theta}(1 - re^{-i\theta})^{-1}$$

$$= \frac{(1 - re^{i\theta})(1 - re^{-i\theta}) + re^{i\theta}(1 - re^{-i\theta}) + re^{-i\theta}(1 - re^{i\theta})}{(1 - re^{i\theta})(1 - re^{-i\theta})}$$

$$= \frac{1-r^2}{1-2r\cos\theta+r^2} \qquad \text{as required.}$$

(ii) We have defined $f(re^{i\theta}) = \tfrac{1}{2}a_0 + \sum_1^\infty r^n(a_n \cos n\theta + b_n \sin n\theta)$ where the

Fourier coefficients $(a_n), (b_n)$ of f_1 are given by the formulae (1.5) of Chapter 1. We thus obtain

$$f(re^{i\theta}) = \frac{1}{2\pi} \int_0^{2\pi} f_1(\phi) \, d\phi + \sum_1^\infty r^n \frac{1}{\pi} \int_0^{2\pi} f_1(\phi)\{\cos n\phi \cos n\theta + \sin n\phi \sin n\theta\} \, d\phi.$$

This is the result of termwise integration in the series

$$\frac{1}{2\pi} \int_0^{2\pi} f_1(\phi)\left\{1 + 2\sum_1^\infty r^n \cos n(\theta - \phi)\right\} d\phi$$

and is thus justified by uniform convergence (Theorem A.27)

It follows that

$$f(re^{i\theta}) = \frac{1}{2\pi} \int_0^{2\pi} f_1(\phi) P_r(\theta - \phi) \, d\phi$$

$$= (f_1 * P_r)(\theta) \qquad \text{as required.}$$

The Poisson kernel $P_r(e^{i\theta})$ has the following important properties.

Lemma 3.5   (i) $\dfrac{1}{2\pi} \displaystyle\int_0^{2\pi} P_r(e^{i\theta}) \, dt = 1$ for $0 \leqslant r < 1$.

(ii) $P_r(r^{i\theta}) > 0$ for $0 \leqslant \theta \leqslant 2\pi$ and $0 \leqslant r < 1$.

(iii) Let δ be a positive real number, $0 < \delta < \pi$, and let $M_r(\delta) = \sup\{P_r(e^{i\theta}): \delta \leqslant \theta \leqslant 2\pi - \delta\}$. Then $M_r(\delta) \to 0$ as $r \to 0$ as $r \to 1_-$.

Proof   (i) is immediate on integrating the expansion

$$P_r(e^{i\theta}) = 1 + 2\sum_1^\infty r^n \cos n\theta$$

(which we know to be uniformly convergent) term by term.

(ii) also follows at once since

$$P_r(e^{i\theta}) = \frac{1 - r^2}{1 - 2r\cos\theta + r^2}.$$

To prove (iii) notice that for fixed r, $P_r(e^{i\theta})$ is an even function of θ which is decreasing on $[0, \pi]$ and increasing on $[\pi, 2\pi]$. Hence $M_r(\delta) = P_r(e^{i\delta})$ which tends to zero as r approaches 1.

Note   A family of functions which has the properties listed in Lemma 3.5 for $(P_r)_{0 \leqslant v < 1}$ is called a *summability kernel*: there are many other examples of such families which are of use in the theory of Fourier series. We have chosen the Poisson kernel because of its simplicity and its importance in applications. Details of other summability kernels can be found in the references cited in the Bibliography.

Now that we have constructed the Poisson integral of f, we must show that it solves the Dirichlet problem.

Theorem 3.6 Let $f_1(e^{i\theta})$ be a real- or complex-valued FC-function on the unit circle, and let $f(re^{i\theta})$ be its Poisson integral. Then (i) f is harmonic in the interior of the unit circle, and (ii) $f(re^{i\theta}) \to f_1(e^{i\theta})$ as $r \to 1_-$ at each point $e^{i\theta}$ at which f_1 is continuous.

Moreover, if f_1 is continuous on a closed subinterval $[\alpha, \beta]$ of values of θ (including continuity on the left at α and on the right at β) then

$$f(re^{i\theta}) \to f_i(e^{i\theta}) \text{ uniformly for } \alpha \leqslant \theta \leqslant \beta.$$

In particular, if f_1 is continuous on the whole unit circle, then $f(re^{i\theta}) \to f_1(e^{i\theta})$ uniformly there.

Proof (i) We know that

$$f(re^{i\theta}) = \frac{1}{2}a_0 + \sum_1^\infty r^n(a_r \cos n\theta + b_n \sin n\theta),$$

where (a_n), (b_n) are the Fourier coefficients of f_1. These coefficients form a bounded sequence (in fact they tend to zero by the Riemann–Lebesgue lemma), and consequently both the given series for f and the series obtained from it by differentiation once or twice with respect to r or θ are uniformly convergent for $0 \leqslant \theta \leqslant 2\pi$, and $0 \leqslant r \leqslant 1 - \delta$ (where δ is an arbitrarily chosen positive number). Consequently, our result on termwise differentiation of series (Theorem A.31) shows that the Laplace operator applied to f may be applied to each term of the above series, which then yields zero since both $r^n \cos n\theta$ and $r^n \sin n\theta$ are harmonic. Hence f is harmonic for $0 \leqslant r \leqslant 1 - \delta$, and since δ was arbitrarily chosen, the result extends to $0 \leqslant r < 1$.

(ii) Let α be a point of continuity of f_1: we shall suppose for simplicity that $\alpha = 0$, otherwise consider $f_1(e^{i(\theta - \alpha)})$. We shall also use the simplified notation mentioned in the note following Definition 3.3 – that is, we shall write $f_1(\theta)$ for $f_1(e^{i\theta})$, etc. We thus obtain

$$f(r) = \frac{1}{2\pi} \int_0^{2\pi} f_1(\theta) P_r(\alpha - \theta) \, d\theta$$

$$= \frac{1}{2\pi} \int_0^{2\pi} f_1(\theta) P_r(\theta) \, d\theta \qquad \text{since } \alpha = 0 \text{ and } P_r(\theta) = P_r(-\theta).$$

We have to show that $f(r) \to f_1(0)$ as $r \to 1_-$.

Suppose $\varepsilon > 0$ is given, and choose $\delta > 0$ so that

$$|f_1(\theta) - f_1(0)| < \varepsilon \qquad \text{if } |\theta| < \delta.$$

We use the periodicity of f_1 and P_r to write

$$f(r) = \frac{1}{2\pi} \int_{-\pi}^\pi f_1(\theta) P_r(\theta) \, d\theta, \qquad \text{and thus}$$

$$f(r) - f_1(0) = \frac{1}{2\pi} \int_{-\pi}^{\pi} [f_1(\theta) - f_1(0)] P_r(\theta) \, d\theta,$$

by Lemma 3.5 (i). We divide this integral into I_1 which is the integral over $[-\delta, \delta]$ and I_2 which is the integral over the rest of $[-\pi, \pi]$.

Then

$$|I_1| \leqslant \frac{1}{2\pi} \int_{-\delta}^{\delta} |f_1(\theta) - f_1(0)| P_r(\theta) \, d\theta \qquad (P_r \text{ is positive})$$

$$\leqslant \varepsilon \frac{1}{2\pi} \int_{-\delta}^{-\delta} P_r(\theta) \, d\theta \leqslant \varepsilon \frac{1}{2\pi} \int_{-\pi}^{\pi} P_r(\theta) \, d\theta = \varepsilon.$$

Having found the above value of δ, we can write

$$|I_2| = \left| \frac{1}{2\pi} \int_{|\theta| \geqslant \delta} [f_1(\theta) - f_1(0)] P_r(\theta) \, d\theta \right| \leqslant \frac{1}{2\pi} K_1 \int_{|\theta| \leqslant \delta} P_r(\theta) \, d\theta$$

where $K_1 = \sup \{|f_1(\theta)| : 0 \leqslant \theta \leqslant 2\theta\}$ is a fixed positive quantity. But $\int_{|\theta| \geqslant \delta} P_r(\theta) \, d\theta \leqslant 2\pi M_r(\delta)$ (see Lemma 3.5(iii)), and this approaches 0 as $r \to 1_-$. Consequently we may choose r so that $M_r(\delta) < \dfrac{\varepsilon}{2K_1}$, and thus $|I_2| < \varepsilon$. It follows that

$$|f(r) - f_1(0)| \leqslant |I_1| + |I_2| < 2\varepsilon, \qquad \text{as required.}$$

Let us now suppose that f_1 is continuous on $[\alpha, \beta]$; it follows that it is uniformly continuous there (Lemma A.4) and we may combine this with the continuity on the left at α and on the right at β to deduce that for any given $\varepsilon > 0$, there is a $\delta > 0$ such that $|f_1(\theta) - f_1(\theta')| < \varepsilon$ whenever $|\theta - \theta'| < \delta$ and one or other of θ, θ' is in $[\alpha, \beta]$.

Suppose that θ is in $[\alpha, \beta]$. We may write

$$f(re^{i\theta}) - f_1(\theta) = \frac{1}{2\pi} \int_{-\pi}^{\pi} (f_1(\phi) - f_1(\theta)) P_r(\theta - \phi) \, d\phi,$$

and divide the range of integration into $[\theta - \delta, \theta + \delta]$ and

$$J_\delta = [-\pi, \pi] \setminus [\theta - \delta, \theta + \delta].$$

This gives firstly $I_1 = (1/2\pi) \int_{\theta-\delta}^{\theta+\delta} (f_1(\phi) - f_1(\theta)) P_r(\theta - \phi) \, d\phi$ and so

$$|I_1| \leqslant \frac{\varepsilon}{2\pi} \int_{\theta-\delta}^{\theta+\delta} P_r(\theta - \phi) \, d\phi \leqslant \varepsilon;$$

and secondly $I_2 = (1/2\pi) \int_{J_\delta} (f_1(\phi) - f_1(\theta)) P_r(\theta - \phi) \, d\phi$ and so

$$|I_2| \leqslant \frac{2K_1}{2\pi} \int_{J_\delta} P_r(\theta - \phi) \, d\phi \leqslant \frac{2K_1}{2\pi} M_r(\delta) \qquad \text{as before.}$$

Thus by making $M_r(\delta) \leqslant \varepsilon\pi/K_1$ we can make $|I_2| \leqslant \varepsilon$ and so $|f(re^{i\theta}) - f_1(\theta)| < 2\varepsilon$.

Since the estimate is independent of the value of θ in $[\alpha, \beta]$, the uniform convergence is established.

This theorem is of fundamental importance for the solution of the Dirichlet problem since it shows that the Poisson integral does have the required boundary values, at least at points of continuity: we state this formally as a corollary.

Corollary 3.7 Let f_1 be a continuous real- or complex-valued function on $[0, 2\pi]$, and let $f(re^{i\delta})$ be its Poisson integral. Then f is harmonic for $0 \leqslant r < 1$, and converges uniformly to f_1 as $r \to 1_-$. Consequently f gives the required harmonic extension of f_1, as required for a solution of the Dirichlet problem.

(Notice also that the *uniqueness* of the solution, which is not guaranteed by the above construction, follows from the result of exercise 15 at the end of the chapter).

Before we consider some examples of the above construction, one final remark about the convergence of $f(re^{i\theta})$ to f_1 is important. The extended function

$$\tilde{f}(re^{i\theta}) = \begin{cases} f(re^{i\theta}) & \text{if } r < 1 \\ f_1(e^{i\theta}) & \text{if } r = 1 \end{cases}$$

is continuous at points on the boundary which are *interior to an interval of continuity* of f_1 (at all points if f_1 is continuous on $[0, 2\pi]$). This means that at such a point, θ_0 say, $f(re^{i\theta}) \to f_1(e^{i\theta_0})$ as $r \to 1$ and $\theta \to \theta_0$; that is we have, in addition to convergence along a radius vector (on which θ is fixed), convergence along *any path from the interior of the disc to the boundary point at* $e^{i\theta_0}$.

The possible existence of radial limits at points of discontinuity is considered in exercise 4.

Examples 3.8 (i) Let f_1 be defined on the unit circle by

$$f_1(e^{i\theta}) = \begin{cases} 1 & \text{for } -\pi/2 < \theta < \pi/2, \\ 0 & \text{elsewhere on } [-\pi, \pi]. \end{cases}$$

The Fourier coefficients of f_1, an even function, are found from

$$a_n = \frac{1}{\pi} \int_{-\pi}^{\pi} f_1(e^{i\theta}) \cos n\theta \, d\theta, \qquad n = 0, 1, 2, \dots$$

$$= \frac{1}{\pi} \int_{-\pi/2}^{\pi/2} \cos n\theta \, d\theta,$$

Thus $a_0 = 1$, while if $n > 0$,

$$a_n = \frac{1}{n\pi} 2 \sin\left(n\frac{\pi}{2}\right) = \begin{cases} 0 & \text{if } n \text{ is even} \\ \dfrac{2}{n\pi}(-1)^{(n-1)/2} & \text{if } n \text{ is odd.} \end{cases}$$

Hence the Poisson integral of f_1 is

$$f(re^{i\theta}) = \tfrac{1}{2}a_0 + \sum_{n=1}^{\infty} a_n r^n \cos n\theta$$

$$= \frac{1}{2} + \frac{2}{\pi} \sum_{k=0}^{\infty} \frac{(-1)^k}{2k+1} r^{2k+1} \cos(2k+1)\theta.$$

putting $n = 2k + 1$.

According to Theorem 3.6, $f(re^{i\theta})$ converges uniformly to $f_1(r^{i\theta})$ on closed intervals of the form $[(-\pi/2) + \delta, (\pi/2) - \delta]$ or $[(\pi/2) + \delta, (3\pi/2) - \delta]$ (though not on $[-\pi/2, \pi/2]$ because of the discontinuities at $\pm \pi/2$). This may be verified for this example as follows. Recall the series for the inverse tangent:

$$\tan^{-1} z = \sum_{k=0}^{\infty} \frac{(-1)^k}{2k+1} z^{2k+1},$$

which is convergent for $|z| \leqslant 1$ expect for $z = \pm 1$.

It follows that for $r < 1$ we can write

$$f(re^{i\theta}) = \frac{1}{2} + \frac{1}{\pi} \sum_{k=0}^{\infty} \frac{(-1)^k}{2k+1} r^{2k+1} \left(e^{i(2k+1)\theta} + e^{-i(2k+1)\theta}\right)$$

$$= \frac{1}{2} + \frac{1}{\pi} \{ \tan^{-1}(re^{i\theta}) + \tan^{-1}(re^{-i\theta}) \}$$

$$= \frac{1}{2} + \frac{1}{\pi} \left\{ \tan^{-1} \left\{ \frac{re^{i\theta} + re^{-i\theta}}{1 - re^{i\theta}re^{-i\theta}} \right\} \right\}$$

$$= \frac{1}{2} + \frac{1}{\pi} \tan^{-1} \left\{ \frac{2r \cos \theta}{1 - r^2} \right\}$$

In particular if $-\pi/2 < \theta < \pi/2$, $\cos \theta > 0$, and $2r \cos \theta / (1 - r^2) \to +\infty$ as $r \to 1_-$, so

$$\frac{1}{2} + \frac{1}{\pi} \tan^{-1} \left\{ \frac{2r \cos \theta}{1 - r^2} \right\} \to \frac{1}{2} + \frac{1}{\pi} \left(\frac{\pi}{2} \right) = 1 \text{ as required.}$$

One can see similarly that the limit is 0 for $(\pi/2) < \theta < (3\pi/2)$, and is $\frac{1}{2}$ if $\theta = \pm \pi/2$ (compare again exercise 4).

This illustrates one of the cases when the Poisson integral can be found explicitly as an elementary function of r and θ.

(ii) The sum of the series for $P_r(\theta)$ found in Lemma 3.4 (i) may be used to give expansions of rational functions of $\cos \theta$. For instance starting from

$$f_1(e^{i\theta}) = \frac{1}{a - \cos \theta},$$

where $a > 1$, we choose a real number r with

$$a = \frac{1}{2r}(1 + r^2) \qquad \text{and } 0 < r < 1:$$

the required value is given by $r^2 - 2ar + 1 = 0$ or $r = a - \sqrt{a^2 - 1}$. We can then write

$$\frac{1}{a - \cos\theta} = \frac{1}{\frac{1}{2r}(1 + r^2) - \cos\theta} = \frac{2r}{1 - 2r\cos\theta + r^2}$$

$$= \frac{2r}{1 - r^2}P_r(\theta) = \frac{2r}{1 - r^2}\left(1 + 2\sum_{n=1}^{\infty} r^n\cos n\theta\right)$$

$$= \frac{1}{\sqrt{a^2 - 1}}\left(1 + 2\sum_{n=1}^{\infty} r^n\cos n\theta\right).$$

For instance, if $a = 3$ then $r = 3 - 2\sqrt{2}$ and we have

$$\frac{1}{3 - \cos\theta} = \frac{1}{2\sqrt{2}}\left(1 + 2\sum_{1}^{\infty} r^n\cos n\theta\right).$$

3.3 APPLICATIONS TO FOURIER SERIES

At this point we shall pause to compare the convergence theory for Fourier series in Chapter 2 (particularly Theorem 2.3) with the results on the Poisson integral which we have obtained so far in this chapter. In particular, we saw that in order to ensure that the Fourier series of a given function converges at some point, something more than mere continuity is required – in our case we have taken Lipschitz continuity. (Also the final section of Chapter 2 and Theorem 2.15 in particular show that this is inevitable: Fourier series of continuous functions sometimes diverge.) On the other hand, Theorem 3.6 has just shown that we obtain radial limits of the Poisson integral of f under less stringent conditions – continuity alone *is* sufficient here.

Thus we have shown that, writing

$$t_n = a_n\cos n\theta + b_n\sin n\theta$$

for the nth term of the Fourier series of f, the series $\sum_{n=0}^{\infty}t_n$ may not converge (i.e. $\sum_{n=0}^{N}t_n$ does not approach a limit as $N \to \infty$) although the 'radial limit',

$\lim_{r\to 1}\sum_{n=0}^{\infty}t_n r^n$, does exist.

In this section we shall investigate the relationship between these concepts for general series, and then for Fourier series.

Definition 3.9 Let $(t_n)_0^{\infty}$ be any sequence of real or complex numbers. We shall say that the series $\sum_0^{\infty}t_n$ is *P-summable*, with sum L if the limit

$$\lim_{r\to 1_-}\sum_0^{\infty}t_n r^n$$

exists and equals L. We may also write $\sum_{n=0}^{\infty}t_n = L(P)$. (The 'P' in 'P-summability'

64

is in honour of Poisson: the name of Abel is equally often found in this context since both mathematicians developed these ideas in the early nineteenth century.)

We have already seen (in a complicated way) with respect to Fourier series that a series may be P-summable but not convergent (apply Theorem 3.6 to Example 2.15(ii)). One can see this more directly by considering the series given by $t_n = (-1)^n$, $n = 0, 1, 2, \ldots$. This is not convergent, since the terms do not even tend to zero. However, for $0 \leqslant r < 1$, $\sum_0^\infty (-1)^n r^n = 1/(1+r)$ which has a limiting value, namely $\frac{1}{2}$ when $r \to 1_-$. Thus the series $\sum_0^\infty (-1)^n$ is divergent, but P-summable with sum $\frac{1}{2}$. This gives us the first half of the following result.

Lemma 3.10   (i) There exist series which are both divergent and P-summable.

(ii) Every convergent series is P-summable, with P-sum equal to its (ordinary) sum.

Proof   (ii) Suppose that $\sum_0^\infty t_n$ is convergent with sum T: i.e. for each $\varepsilon > 0$, there is some N such that $|\sum_{n=0}^m t_n - T| < \varepsilon$ if $m \geqslant N$.

Now consider $\phi(r) = \sum_0^\infty t_n r^n$ for $0 \leqslant r < 1$: ϕ is well defined since $t_n \to 0$ as $n \to \infty$ and so the series is absolutely convergent by comparison with $\sum_0^\infty r^n$. Our aim is of course to show that $\phi(r) \to T$ as $r \to 1_-$.

To show this, we write $s_m = \sum_0^m t_n$, so that $|s_m - T| < \varepsilon$ if $m \geqslant N$. Then we can write

$$\phi(r) = s_0 + \sum_1^\infty (s_n - s_{n-1}) r^n \qquad (\text{since } s_n - s_{n-1} = t_n \text{ for } n \geqslant 1)$$

$$= \sum_0^\infty s_n (r^n - r^{n+1}) = (1-r) \sum_0^\infty s_n r^n.$$

We also know that $(1-r)\sum_0^\infty r^n = 1$ for $0 \leqslant r < 1$, so that we can write

$$\phi(r) - T = (1-r) \sum_0^\infty (s_n - T) r^n.$$

We now split this summation into two parts, for $0 \leqslant n \leqslant N-1$ and for $n \geqslant N$. For the second summation we have

$$\left| (1-r) \sum_N^\infty (s_n - T) r^n \right| \leqslant (1-r) \sum_N^\infty \varepsilon r^n = \varepsilon r^N < \varepsilon.$$

For the first summation, suppose S is an upper bound for the sequence $(|s_n|)$, so that

$$\left| (1-r) \sum_0^{N-1} (s_n - T) r^n \right| \leqslant (1-r) \sum_0^{N-1} (S + |T|) r^n$$

$$= (1-r)(S + |T|)N.$$

This too can be made less than ε if we choose $0 < 1 - r < \varepsilon\{(S+|T|)N\}^{-1} = \delta$, say. Thus we have shown that

$$|\phi(r) - T| < 2\varepsilon \qquad \text{if } 1 - \delta < r < 1, \text{ or that}$$
$$\phi(r) \to T \text{ as } r \to 1_-, \text{ as required.}$$

The result we have just proved shows both that P-summability is compatible with ordinary convergence (that is, we cannot use it to get a different answer to the ordinary sum of a series) but also that it includes convergence: every convergent series is P-summable to the same sum.

Example 3.11 Find the P-sum of the (divergent) series

$$1 - 2 + 3 - 4 + \cdots = \sum_0^\infty (-1)^n (n + 1).$$

In this example, $t_n = (-1)^n (n + 1)$ and so we must consider

$$\phi(r) = \sum_0^\infty t_n r^n = \sum_0^\infty (n + 1)(-r)^n.$$

But this series is the (binomial) expansion of $(1 + r)^{-2}$ which is convergent for $|r| < 1$, so $\phi(r) = (1 + r)^{-2}$ for $0 \leqslant r < 1$.
   In particular, $\phi(r) \to \frac{1}{4}$ as $r \to 1_-$, so the series is P-summable, and $\sum_0^\infty (-1)^n (n + 1) = \frac{1}{4}(P)$.

This striking example shows that a P-summable series may have terms which are unbounded. It should not be thought, however, that all series are P-summable. For instance, if $t_n = 1$ for all n,

$$\phi(r) = \frac{1}{1 - r} \to \infty \qquad \text{as } r \to 1_-,$$

while if $t_n = (-2)^n$, $\phi(r) = \sum_0^\infty (-2r)^n$ is only convergent for $|r| < \frac{1}{2}$ and so cannot have a limit as $r \to 1_-$.
   We can now restate the important Theorem 3.6 in the language of P-summability.

Corollary 3.12 Let f be an FC-function on $[0, 2\pi]$, whose Fourier series is $S(f) = \frac{1}{2}a_0 + \sum_1^\infty (a_n \cos nx + b_n \sin nx) = \sum_{-\infty}^\infty c_n e^{inx}$.
   Then $S(f)$ is P-summable, with sum $f(x_0)$, at any point x_0 of continuity of f. The limit is attained uniformly on any closed interval on which f is continuous, and in particular if f is continuous on the whole of $[0, 2\pi]$.

It follows from this corollary, and the compatibility of P-summability with ordinary convergence, that if $S(f)$ converges at a point x_0 at which f is continuous, then its sum must equal $f(x_0)$ there.
   Notice that exercise 4 at the end of this chapter may be restated to give information about the P-summability of Fourier series at points of discontinuity. Corollary 3.12 also provides us with a new proof of Theorem 1.10. For if all the

Fourier coefficients of f are zero, then

$$f(r,x) = \frac{1}{2}a_0 + \sum_1^\infty r^n(a_n \cos nx + b_n \sin nx)$$

vanishes identically, and the Fourier series of f is thus P-summable to zero everywhere. But the P-sum is equal to f at points of continuity, so f must vanish at all points of continuity.

Finally it should be pointed out that P-summability is only one of a large number of possible summability processes which can be used to extend the notation of sum to certain non-convergent series. We have chosen P-summability because of its natural relation to Fourier series via the Poisson integral. For an account of other processes the reader can consult Zygmund (1959) for applications to trigonometric series, or Hardy (1963) for the general theory.

3.4 HARMONIC CONJUGATES

We return to the subject of harmonic functions in the unit disc, to consider a topic which was suggested in Section 3.1, namely the relation between a real-valued harmonic function u and an analytic function f of which u is the real part. In Section 3.1 this relationship was considered only for a few elementary functions, namely powers and the exponential and logarithmic functions: we shall now investigate the relationship in more generality. The relationship is also important in physical applications where the relation between the harmonic function $u = \text{Re } f$ and $v = \text{Im } f$ carries information about lines of force in magnetic or gravitational fields: we shall look into this in the final section of the chapter.

We begin by showing that the method suggested in Section 3.1 for showing that the real and imaginary parts of an analytic function are harmonic, works equally well in the general case. An *analytic function f* of a complex variable is one which has a complex derivative

$$f'(z) = \lim_{n \to 0} \frac{1}{h}(f(z + h) - f(z))$$

at all points of its domain of definition (the limit is taken through complex values of h naturally), and we shall assume it known that this entails the existence of all higher derivatives. In particular, if we write $f(z) = f(x + iy) = u(x, y) + iv(x, y)$ then $u = \text{Re } f$ and $v = \text{Im } f$ have continuous partial derivatives of all orders.

Lemma 3.13 Let f be analytic on an open set G in $\mathbb{R}^2$. Then $u = \text{Re } f$ and $v = \text{Im } f$ are harmonic on G.

Proof We write $f(x + iy) = u(x, y) + iv(x, y)$ and differentiate partially with respect to x and y (compare Example 3.2(i)).

We get

$$f'(x + iy) = \frac{\partial u}{\partial x}(x, y) + i\frac{\partial v}{\partial x}(x, y),$$

and

$$if'(x + iy) = \frac{\partial u}{\partial y}(x, y) + i\frac{\partial v}{\partial y}(x, y).$$

It follows that

$$i\left(\frac{\partial u}{\partial x} + i\frac{\partial v}{\partial x}\right) = \frac{\partial u}{\partial y} + i\frac{\partial y}{\partial y},$$

or

$$\frac{\partial u}{\partial x} = \frac{\partial v}{\partial y} \quad \text{and} \quad -\frac{\partial v}{\partial x} = \frac{\partial u}{\partial y}$$

on equating real and imaginary parts.

However, we know that all partial derivatives of u and v of all orders exist and are continuous: consequently all mixed partial derivatives are equal, so that we may write

$$\frac{\partial^2 u}{\partial x^2} = \frac{\partial}{\partial x}\left(\frac{\partial v}{\partial y}\right) = \frac{\partial}{\partial y}\left(\frac{\partial v}{\partial x}\right) = -\frac{\partial^2 u}{\partial y^2}.$$

Thus u is harmonic, and a similar argument applies to v.

Example 3.14   Let $f(z) = \sqrt{z}$, where for definiteness we shall put $z = x + iy$ and restrict ourselves to the region G where $y > 0$, and take the value of the square root which lies in the first quadrant. (Equivalently, let $z = re^{i\theta}$, where $0 < \theta < \pi$, and then take $\sqrt{r}\, e^{\frac{1}{2}i\theta}$ for $\sqrt{z}$).

Then if $z = u + iv$, we can find u and v as follows.

Since $(u + iv)^2 = z = x + iy$ we have

$$u^2 - v^2 = x,$$
$$2uv = y,$$

and so $(u^2 + v^2)^2 = x^2 + y^2 = r^2$, say.

It follows that $u^2 + v^2 = r$, and so $2u^2 = r + x$, $2v^2 = r - x$. It follows that $f(z) = u + iv = \sqrt{\frac{1}{2}(r + x)} + i\sqrt{\frac{1}{2}(r - x)}$, where the square roots are now the (non-negative) roots of (non-negative) real numbers. Hence (discarding the numerical factors) we have found new harmonic functions $\sqrt{r \pm x}$ as real and imaginary parts of an analytic function. The alternative method of verifying that $\sqrt{r \pm x}$ gives solutions of Laplace's equation is a tedious exercise in partial differentiation which we omit.

Our main objective in this section is the reverse of the above process, namely beginning with a harmonic function u in the unit disc (or occasionally a more general region), to find an analytic function f of which u is the real part. We shall see that f always exists when u is defined on a disc (and often in more general situations): when it does exist it is uniquely defined except for an arbitrary additive constant. The existence of such a function f of course entails the existence of $v = \text{Im } f$, and it is on v that we shall concentrate from now on.

Definition 3.15 Let u be harmonic on an open set G in $\mathbb{R}^2$. If a function v exists such that $f = u + iv$ is analytic on G then we say that v is a *harmonic conjugate* of u.

Our next result establishes the existence and uniqueness of v.

Theorem 3.16 Let u be harmonic in a disc D with centre (x_0, y_0). Then there exists a harmonic function v for which $f = u + iv$ is analytic on D. The function v is determined uniquely to within an additive constant.

Proof We notice first that since v must satisfy the Cauchy–Riemann equations

$$\frac{\partial u}{\partial x} = \frac{\partial v}{\partial y}, \qquad \frac{\partial u}{\partial y} = -\frac{\partial v}{\partial x},$$

then any two harmonic conjugates must have the same partial derivatives with respect to both x and y, and so their difference must be constant along lines where x or y are constant: it follows at once that the difference of two harmonic conjugates must be constant, which establishes the uniqueness assertion in the theorem.

To show existence, we display an explicit construction for v: the motivation for the formula comes from complex integration and can be found as exercise 7. In the formulae we denote the partial derivatives of u (as functions) by $\partial u / \partial x$, $\partial u / \partial y$, and their values at particular points by $\partial u/\partial x(x, y)$, $\partial u/\partial y(t, y_0)$, etc.

We define for any point (x, y) in D,

$$v(x, y) = -\int_{x_0}^{x} \frac{\partial u}{\partial y}(t, y_0)\, dt + \int_{y_0}^{y} \frac{\partial u}{\partial x}(x, u)\, du.$$

Figure 3.1 shows the paths of integration used from (x_0, y_0) to (x, y) in the case when $x > x_0$, $y > y_0$.

Notice that in the first integral the value of the second co-ordinate has the fixed value y_0, and in the second integral the first co-ordinate is equal to x.

It is easy to find the partial derivative of v with respect to y since y occurs only as an upper limit. Hence we obtain

$$\frac{\partial v}{\partial y}(x, y) = \frac{\partial u}{\partial x}(x, y), \qquad \text{as required.}$$

To find the x-derivative we must take into account both the first occurrence of x as an upper limit and its second occurrence inside an integral. Using the formula for differentiation under the integral sign, we obtain

$$\frac{\partial v}{\partial x} = -\frac{\partial u}{\partial y}(x, y_0) + \int_{y_0}^{y} \frac{\partial^2 u}{\partial x^2}(x, u)\, du.$$

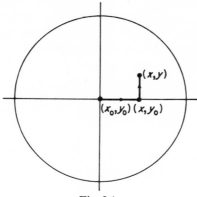

Fig. 3.1.

Since u is harmonic we have $\partial^2 u/\partial x^2 = -\partial^2 u/\partial y^2$ and we obtain successively

$$\frac{\partial v}{\partial x} = -\frac{\partial u}{\partial y}(x, y_0) - \int_{y_0}^{y} \frac{\partial^2 u}{\partial y^2}(x, u)\,\mathrm{d}u$$

$$= -\frac{\partial u}{\partial y}(x, y_0) - \left[\frac{\partial u}{\partial y}(x, u)\right]_{u=y_0}^{u=y}$$

$$= -\frac{\partial u}{\partial y}(x, y_0) - \left[\frac{\partial u}{\partial y}(x, y) - \frac{\partial u}{\partial y}(x, y_0)\right]$$

$$= -\frac{\partial u}{\partial y}(x, y).$$

This shows that the function v which we have constructed does indeed satisfy the Cauchy–Riemann equations. The proof that $f = u + iv$ is complex differentiable follows easily from this and the definition of differentiability of functions of two variables, and is omitted.

The theorem we have just proved gives us an extra piece of information, namely a method for the explicit structure of harmonic conjugates, as the first of the following examples shows.

Examples 3.17 (i) Let $u(x, y) = xy$. Since $\partial^2 u/\partial x^2 = \partial^2 u/\partial y^2 = 0$, u is harmonic everywhere. We construct its harmonic conjugate using the method of the proof of Theorem 3.16. We have $\partial u/\partial x = y$, $\partial u/\partial y = x$, so that $\partial u/\partial y(t, y_0) = t$, $\partial u/\partial x(x, u) = u$. Hence (putting $x_0 = y_0 = 0$) the formula gives

$$v(x, y) = -\int_0^x t\,\mathrm{d}t + \int_0^y u\,\mathrm{d}u = -\tfrac{1}{2}(x^2 - y^2).$$

(ii) If v is the harmonic conjugate of u, then $-u$ is the harmonic conjugate of v. This follows either from the Cauchy–Riemann equations, or by observing that if $f = u + iv$ is analytic, so is $-if = v - iu$.

(iii) If $u = r^n \cos n\theta$, $n = 0, 1, 2, 3, \ldots$, in polar co-ordinates, then its harmonic conjugate is $r^n \sin n\theta$ (see the discussion following Example 3.2).

(iv) Suppose $u = P_r(\theta) = 1 + 2\sum_{n=1}^{\infty} r^n \cos n\theta$, the Poisson kernel.

The harmonic conjugate of u, which we shall denote by $Q(\theta)$, is called the *conjugate Poisson kernel*:

$$Q_r(\theta) = 2\sum_{n=1}^{\infty} r^n \sin n\theta, \qquad 0 \leqslant r < 1.$$

To find the sum of the series we can either write it as

$$-i\sum_{n=1}^{\infty} r^n(e^{in\theta} - e^{-in\theta})$$

and sum the geometric series (exercise 8), or notice that

$$u = \mathrm{Re}\left(1 + 2\sum_{1}^{\infty} z^n\right) \qquad \text{where } z = re^{i\theta}$$

$$= \mathrm{Re}\left(1 + \frac{2z}{1-z}\right) = \mathrm{Re}\left(\frac{1+z}{1-z}\right).$$

We can write

$$\frac{1+z}{1-z} = \frac{1+re^{i\theta}}{1-re^{i\theta}} = \frac{(1+re^{i\theta})(1-re^{-i\theta})}{1-2r\cos\theta + r^2}$$

$$= \frac{1-r^2 + 2ir\sin\theta}{1-2r\cos\theta + r^2}.$$

This recovers the known formula for $u = P_r(\theta)$, and also gives

$$v = Q_r(\theta) = \frac{2r\sin\theta}{1-2r\cos\theta + r^2}.$$

Example 3.17(iv) suggests a quite different approach to the subject of harmonic conjugates. In situations where a harmonic function u is known to be given in terms of its boundary values by the Poisson interval,

$$u(re^{i\theta}) = f_1 * P_r(\theta)$$

we can now consider the representation of its harmonic conjugate
$$v(re^{i\theta}) = f_1 * Q_r(\theta).$$

The consistency of this approach is guaranteed by the observation in 3.17(iii). We shall investigate this aspect of the relation between a function and its harmonic conjugate more thoroughly in Chapter 4.

Unfortunately, in an arbitrary domain, the existence of a harmonic conjugate is not always assured. The standard example of this is as follows.

Example 3.18 The function $u(x, y) = \log r (r = \sqrt{x^2 + y^2})$ is harmonic on $\mathbb{R}^2 \backslash (0, 0)$ but has no harmonic conjugate there.

For if we differentiate, we obtain

$$\frac{\partial u}{\partial x} = \frac{1}{r}\frac{\partial r}{\partial x} = \frac{1}{r}\frac{x}{r} = \frac{x}{r^2}, \qquad \frac{\partial^2 u}{\partial x^2} = \frac{r^2 - x2rx/r}{r^4} = \frac{r^2 - 2x^2}{r^4}.$$

Similarly,

$$\frac{\partial^2 u}{\partial y^2} = \frac{r^2 - 2y^2}{r^2} \qquad \text{(notice that } r > 0 \text{ throughout) and hence}$$

$$\frac{\partial^2 u}{\partial x^2} + \frac{\partial^2 u}{\partial y^2} = \frac{2r^2 - 2x^2 - 2y^2}{r^2} = 0,$$

which proves that u is harmonic. However, if we suppose that v is a harmonic conjugate of u, we must have $f = u + iv$ analytic in $\mathbb{R}^2 \backslash (0, 0)$.

Now $u = \operatorname{Re} \log z$ (except on the negative x-axis where $\log z$ is undefined), and so v must coincide with $\operatorname{Arg} z$. However, $\operatorname{Arg} z$ has a discontinuity across the negative x-axis and so v cannot be defined on the whole of $\mathbb{R}^2 \backslash (0, 0)$. Notice that $v = \operatorname{Arg} z$ *does* supply a harmonic conjugate on the region obtained by deleting the whole of the negative imaginary axis.

In the next section we shall see that for a reasonably large class of regions, harmonic conjugates do exist: this will follow from our solution of the Dirichlet problem in such regions.

3.5 THE DIRICHLET PROBLEM IN SOME OTHER REGIONS

This section is concerned with the problem of extending the results of the earlier part of the chapter, in which we solved the Dirichlet problem for a disc, to other regions. That is, we shall consider the situation in which a function f is given on the boundary of some region, and we want to find a harmonic function u on the region, which has f as boundary values. The regions we consider will not be very general in nature – in fact we shall consider examples whose boundary consists of a finite number of lines or circular arcs. The theory in the general case is very deep (see for instance Ahlfors, 1966) and uses methods which are quite beyond our scope.

Our technique will be to find a mapping which maps the given region into a circle, to solve the problem thus obtained using our known results, and then to convert this solution into a solution of the original problem by the reverse mapping. For this method to be successful, it is obviously necessary that the mapping used should preserve the class of harmonic functions: if ϕ maps $R_1 \rightarrow R_2$ then we need to know that $u(\phi)$ is harmonic on R_1 whenever u is harmonic on R_2. The required mappings are the analytic functions and we begin by establishing this point.

Theorem 3.19 If ϕ is an analytic map from the region R_1 onto the region R_2 and

u is harmonic on R_2, then $u_1 = u(\phi)$ is harmonic on R_1. Conversely, if u_1 is harmonic whenever u is, then ϕ must either be analytic or the complex conjugate of an analytic function.

Proof Let ϕ be analytic from R_1 to R_2 and u be harmonic on R_2: we can suppose without loss of generality that u is real valued. Then it follows from Theorem 3.16 that if we specify some point of R_2 then there exists in a neighbourhood N of that point, an analytic function f with $u = \operatorname{Re} f$. Then $f \circ \phi$ is analytic on any subset of R_1 which maps into N, and thus $u_1 = \operatorname{Re} f(\phi)$ is harmonic on any such subset. It follows that since ϕ is defined on the whole of R_1, then u_1 is harmonic throughout R_1 also.

Conversely, let us suppose that $u_1 = u(\phi)$ is harmonic whenever u is. If we take the special cases $u(z) = \operatorname{Re} z$, $\operatorname{Im} z$, it follows that ϕ is equal to $\phi_1 + i\phi_2$ where ϕ_1 and ϕ_2 are harmonic, and it remains to establish the relationship between ϕ_1 and ϕ_2, i.e. that $\phi_1 \pm i\phi_2$ is analytic.

Let us now take $u(z) = \operatorname{Re}(z^2), \operatorname{Im}(z^2)$: it follows that both $\phi_1^2 - \phi_2^2$ and $\phi_1\phi_2$ are harmonic, and hence if we denote partial differentiation with respect to x, y by D_1, D_2 respectively, we have successively.

$$D_1(\phi_1\phi_2) = (D_1\phi_1)\phi_2 + \phi_1 D_1\phi_2,$$
$$D_{11}(\phi_1\phi_2) = (D_{11}\phi_1)\phi_2 + 2D_1\phi_1 D_1\phi_2 + \phi_1 D_{11}\phi_2$$
$$D_{22}(\phi_1\phi_2) = (D_{22}\phi_1)\phi_2 + 2D_2\phi_1 D_2\phi_2 + \phi_1 D_{22}\phi_2$$

and hence $D_1\phi_1 D_1\phi_2 + D_2\phi_1 D_2\phi_2 = 0$, using the harmonic property of ϕ_1, ϕ_2 and $\phi_1\phi_2$.

A similar calculation applied to $\phi_1^2 - \phi_2^2$ shows that

$$(D_1\phi_1)^2 - (D_1\phi_2)^2 + (D_2\phi_1)^2 - (D_2\phi_2)^2 = 0.$$

We write for simplicity $a = D_1\phi_1$, $b = D_1\phi_2$, $c = D_2\phi_1$, $d = D_2\phi_2$, so that these equations become

$$ab + cd = 0,$$
$$a^2 - b^2 + c^2 - d^2 = 0, \text{ respectively.}$$

Multiply the latter by a^2 and substitute for a^2b^2 to obtain

$$a^2 - c^2 d^2 + a^2(c^2 - d^2) = 0,$$

or

$$(a^2 + c^2)(a^2 - d^2) = 0.$$

Hence either $a = c = 0$ in which case $b = d = 0$ also, and ϕ is a constant mapping, or $a^2 = d^2$, in which case $b^2 = c^2$ also. In the latter case we have $a = \varepsilon d, b = -\varepsilon c$, where $\varepsilon = \pm 1$. If $\varepsilon = 1$ these are the Cauchy–Riemann equations and ϕ is analytic, while if $\varepsilon = -1$, it follows similarly that $\bar{\phi}$ is analytic.

Our next task is to consider some examples of analytic mappings and their effect on some particular regions. We shall be particularly interested in bilinear mappings, that is in functions of the form

$$w = f(z) = \frac{az + b}{cz + d}$$

where a, b, c, d are complex constants, and where in order, for f to be non-constant we require that $ad - bc \neq 0$. Particular examples of such mappings are:

(a) *dilation*: if $a/d = r$ is real and positive and $b = c = 0$, then $f(z) = rz$, which changes all distances from the origin by a fixed ratio r;
(b) *rotation*: if $a/d = e^{i\theta}$ for some real θ, and $b = c = 0$, then $f(z) = e^{i\theta}z$, which rotates all points about the origin through an angle θ;
(c) *translation*: if $a = d \neq 0$, $c = 0$, then $f(z) = z + b/d$, which translates all points through a fixed amount;
(d) *inversion*: if $a = d = 0$, $b = c \neq 0$, then $f(z) = 1/z$, which maps each point $z = re^{i\theta}$ to $1/re^{-i\theta}$, which is the reflection of its inverse in the unit circle (Fig. 3.2).

Notice that c and d cannot both vanish, so that if $c \neq 0$ we can write

$$f(z) = \frac{az + b}{cz + c} = \frac{a}{c} - \frac{ad - bc}{c}(cz + d)^{-1},$$

while if $c = 0$, we have

$$f(z) = \frac{a}{d}z + \frac{b}{d};$$

in either case this shows that f can be written as a composition of the above four special types of bilinear mapping.

The importance of bilinear mappings is that they are one-to-one on the extended complex plane (including the point at infinity) while preserving the class of all circles and straight lines.

The class of bilinear mappings is in fact characterized by these properties, though we shall not prove this. For a proof, and a deeper discussion of properties of bilinear mappings, see Ahlfors (1966).

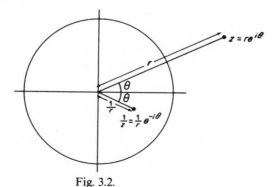

Fig. 3.2.

Theorem 3.20 (i) A curve in the complex plane is a circle or a straight line if and only if its equation can be put into the form

$$A z\bar{z} + B z + \bar{B}\bar{z} + C = 0$$

for suitable constants A, B, C, with A and C real.

(ii) The class of all circles and straight lines is preserved by any bilinear mapping.

Proof (i) This is omitted: see exercise 2.

(ii) Let the mapping be given by

$$w = \frac{az + b}{cz + d}, \text{ with } ad - bc + 0:$$

then the inverse mapping is given by

$$z = \frac{-dw + b}{cw - a}.$$

If we substitute this into

$$A z\bar{z} + B z + \bar{B}\bar{z} + C = 0, \text{ we obtain}$$
$$A_1 w\bar{w} + B_1 w + \bar{B}_1 \bar{w} + C_1 = 0,$$

where

$$A_1 = Ad\bar{d} - Bd\bar{c} - \bar{B}\bar{d}c + Cc\bar{c}, \text{ is real,}$$
$$B_1 = -Ad\bar{b} - Bd\bar{a} + \bar{B}\bar{b}c - Cc\bar{a},$$

and

$$C_1 = Ab\bar{b} - Bb\bar{a} - \bar{B}\bar{b}a + Ca\bar{a} \text{ is also real.}$$

Hence the image is again a circle or straight line, and this completes the proof.

We now consider some particular mappings which will be useful in the applications to come. Notice that, as a general principle, any circle or straight line which passes through the point $-d/c$ (which is mapped by f to the point at infinity) is mapped to a straight line, while any circle or straight line not through $-d/c$ is mapped to a circle.

Examples 3.21 (i) The inversion $f(z) = 1/z$ maps any circle or straight line through the origin to a straight line, as noted above. In particular, the circle C with centre at $z = 1$ with radius 1, passes through $z = 0$ and $z = 2$, and so is mapped to a straight line through $z = \frac{1}{2}$: f also maps conjugate numbers to conjugate values ($f(\bar{z}) = \overline{f(z)}$), so in fact the image must be the line given by $Re\, z = \frac{1}{2}$.

(ii) The mapping

$$f(z) = \frac{1-z}{1+z}$$

maps the imaginary axis into the unit circle, $|z| = 1$, since for real y,

$$\left|\frac{1-iy}{1+iy}\right| = 1.$$

The mapping also has the useful property of being its own inverse, since $f(f(z)) = z$, and consequently it maps the unit circle to the imaginary axis, as one can verify independently. Hence the half-planes $\operatorname{Re} z > 0$ and $\operatorname{Re} z < 0$ are mapped into the interior and exterior of the unit circle (respectively, since $f(1) = 0$).

(iii) For any complex number a, $|a| < 1$, the mapping

$$f(z) = \frac{z-a}{1-\bar{a}z}$$

takes a to the origin while keeping the unit circle invariant: for if $z = e^{i\theta}$,

$$|z-a| = |e^{i\theta} - a| = |1 - ae^{-i\theta}| < |1 - \bar{a}e^{i\theta}| = |1 - \bar{a}z|.$$

We next give a couple of examples of other functions which can sometimes be combined with linear mappings.

Examples 3.22 (i) (Mapping of circular sectors) The mapping $f(z) = z^n$ ($n = 1, 2, 3, \ldots$) is analytic everywhere, and maps the sector $\{z: 0 < \operatorname{Arg} z < \pi/n\}$ into the upper half-plane $\{z: \operatorname{Im} z > 0\}$.

(ii) (Mapping of half or whole strips) The mapping $f(z) = \log z$ maps the semicircle $\{z: |z| < 1, \operatorname{Im} z > 0\}$ onto the half-strip $\{z: \operatorname{Re} z < 0, 0 < \operatorname{Im} z < \pi\}$, and the upper half-plane $\{z: \operatorname{Im} z > 0\}$ onto the whole strip $\{z: 0 < \operatorname{Im} z < \pi\}$.

(iii) (Mapping a circle onto an equilateral triangle) Consider the mapping

$$f(z) = \int_0^z \frac{du}{(1-u^3)^{2/3}}$$

where for $|z| \leqslant 1$ the integral is taken along the straight line from the origin to the point z. For u on this line segment we also have $|u| \leqslant 1$ and so $\operatorname{Re}(1 - u^3) \geqslant 0$: consequently the fractional power can be given its principal value and the integral is well defined. In particular,

$$f(1) = \int_0^1 \frac{du}{(1-u^3)^{2/3}} = \frac{2}{3}\int_0^{\pi/2} \frac{(\sin\theta)^{-1/3}\cos\theta}{(\cos^2\theta)^{2/3}} d\theta \qquad \text{on putting } u = (\sin\theta)^{2/3}$$

$$= \frac{2}{3}\int_0^{\pi/2} \frac{d\theta}{(\sin\theta\cos\theta)^{1/3}} = \frac{1}{3}\frac{(\Gamma(\tfrac{1}{3}))^2}{\Gamma(\tfrac{2}{3})} = 1.7666388\ldots$$

where Γ is Euler's gamma function. This constant we shall denote by α in what follows.

To show that the image of the unit circle is actually an equilateral triangle, consider first the effect which f has on the point 1, ω, ω^2, where as usual $\omega = -\frac{1}{2} + i\sqrt{3}/2 = e^{2\pi i/3}$. We have already seen that $f(1) = \alpha$, and it follows that

$$f(\omega) = \int_0^\omega \frac{du}{(1-u^3)^{3/2}} = \omega\alpha \qquad \text{on putting } u = \omega v,$$

and similarly

$$f(\omega^2) = \omega^2\alpha.$$

To find the effect which f has on other points of the unit circle, suppose first that $z = e^{i\theta}$, $0 < \theta < 2\pi/3$.

Then

$$f(e^{i\theta}) = \int_0^{e^{i\theta}} \frac{du}{(1-u^3)^{2/3}},$$

and if we differentiate this with respect to θ we obtain

$$\frac{ie^{i\theta}}{(1-e^{3i\theta})^{2/3}} = g(\theta), \text{ say.}$$

To simplify this expression, we write

$$1 - e^{3i\theta} = -e^{3/2i\theta}\, 2i \sin\left(\tfrac{3}{2}\theta\right)$$

$$= 2e^{i(3\theta/2 - \pi/2)} \sin\left(\tfrac{3}{2}\theta\right),$$

and notice that since $0 < \theta < 2\pi/3$, $\sin\left(\tfrac{3}{2}\theta\right)$ is *positive*.

Hence the principal value of $(1 - e^{3i\theta})^{2/3}$ is $2^{2/3} e^{i(\theta - \pi/3)} \left(\sin\tfrac{3}{2}\theta\right)^{2/3}$, and so

$$\frac{d}{d\theta}\{f(e^{i\theta})\} = g(\theta) = 2^{-2/3}\, ie^{i\pi/3} (\sin\tfrac{3}{2}\theta)^{2/3}$$

$$= e^{i5\pi/6} r(\theta) \text{ say, where } r(\theta) > 0.$$

It follows that as θ increase from 0 to $2\pi/3$, $f(e^{i\theta})$ moves from $z = \alpha$ to $z = \omega\alpha$ in the fixed direction at an angle $5\pi/6$ to the real axis, so the image of the circular arc from 1 to ω is the straight line from α to $\omega\alpha$.

Similar calculations show that the arcs from ω to ω^2 and ω^2 to 1 are mapped onto the line segments from $\omega\alpha$ to $\omega^2\alpha$, and $\omega^2\alpha$ to α, which completes the argument (see Fig. 3.3).

A more general result shows that

$$f(z) = \int_0^z \frac{du}{(1-u^n)^{2/n}}$$

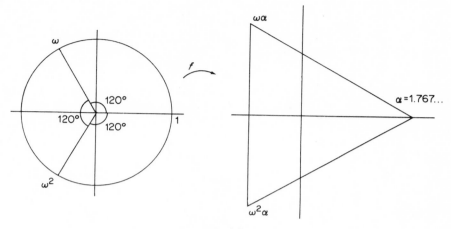

Fig. 3.3.

maps a circle onto a regular n-sided polygon: in particular,

$$f(z) = \int_0^z \frac{du}{(1 - u^4)^{1/2}}$$

maps a circle to a square

We complete this section with a couple of example which show how analytic mapping may be used to solve the Dirichlet problem in some non-circular regions.

Example 3.23 Consider the semi-circular region R_1 given by $\{z: \operatorname{Re} z \geqslant 0, |z| \leqslant 1\}$, and suppose that a function is given which takes the value 1 on the circular arc of the boundary, and 0 on the interval $[-i, i]$. We shall find the harmonic function f which has ϕ for boundary values, and in particular, find $f(\tfrac{1}{2})$.

We solve this problem in two stages. The first consists of finding a mapping of the semicircular region R_1 onto the unit circle. To do this we begin with a bilinear mapping ϕ_1 which maps R_1 onto the first quadrant,

$$R_2 = \{z: \operatorname{Re} z > 0, \operatorname{Im} z > 0\}.$$

We take

$$\phi_1(z) = -i\left(\frac{z+i}{z-i}\right)$$

since this takes $-i, i, 0, 1$ into $0, \infty, i, 1$ respectively, and the points on the segment $(0, 1)$ onto the segment of the unit circle in the first quadrant, (Fig. 3.4).

The mapping $\phi_2(z) = -iz^2$ then maps R_2 to the right-hand half-plane $\{z: \operatorname{Re} z > 0\}$, and finally

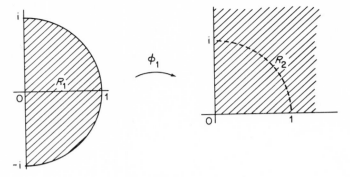

Fig. 3.4.

$$\phi_3(z) = \frac{1-z}{1+z} \qquad \text{(Example 3.21(ii))}$$

takes this to the unit disc.

The required mapping is then the composite $\phi_3(\phi_2(\phi_1(z)))$ which one can easily check simplifies to give

$$1 - i\left(\frac{z+i}{z-i}\right)^2 \bigg/ 1 + i\left(\frac{z+i}{z-i}\right)^2 = \frac{-i(z^2 + 2z - 1)}{z^2 - 2z - 1}.$$

The factor $-i$ is plainly unnecessary since it only rotates the unit disc, and we may take our required mapping in the form

$$\phi(z) = \frac{2z - (1 - z^2)}{2z + (1 - z^2)}.$$

(As a check, notice that $\phi(0) = -1, \phi(1) = 1, \phi(i) = i, \phi(-i) = -i$.)

If we now transfer the boundary values, we see that we have to solve the Dirichlet problem in the unit disc D, with boundary values $+1$ on the boundary with $\mathrm{Re}\, z \geqslant 0$, and 0 on the rest of the boundary. This means that we have to find the Fourier coefficients $(a_n)_0^\infty$ of the function u which is equal to 1 on $[-\pi/2, \pi/2]$ and zero elsewhere, and then give the solution in the form $\frac{1}{2}a_0 + \sum_{n=1}^\infty a_n r^n \cos n\theta$ (the function u is even, so there are no sine terms in the expansion). The required Fourier coefficients were found in Example 3.8(i) to be given by

$$a_n = \begin{cases} 0 & \text{if } n \text{ is even and } \geqslant 2, \\ (-1)^{(n-1)/2} & \text{if } n \text{ is odd.} \end{cases}$$

Hence the function in the unit disc with the given boundary values is

$$u(r, \theta) = \frac{1}{2} + \frac{2}{\pi}\sum_0^\infty \frac{(-1)^k}{(2k+1)} r^{2k+1} \cos(2k+1)\theta$$

or

$$u(w) = \frac{1}{2} + \frac{1}{\pi}\sum_0^\infty \frac{(-1)^k}{2k+1}(w^{2k+1} + \bar{w}^{2k+1}), \text{ where } w = re^{i\theta}.$$

To solve the Dirichlet problem in the original semicircular region we transfer this solution via the mapping $w = \phi(z)$ found above, so that the required solution in R_1 is given by

$$f(z) = u(\phi(z)) = \frac{1}{2} + \frac{1}{\pi}\sum_0^\infty \frac{(-1)^k}{2k+1}\{(\phi(z))^{2k+1} + (\overline{\phi(z)})^{2k+1}\}.$$

In particular, $\phi(\frac{1}{2}) = \frac{1}{7}$, so $f(\frac{1}{2}) = u(\frac{1}{7}) = \frac{1}{2} + (2/\pi)\sum_0^\infty (-1)^k/(2k+1)7^{-2k-1}$. (Both this value and the general solution may be further simplified if needed by using the series expansion for the inverse tangent: for instance, $f(\frac{1}{2}) = \frac{1}{2} + 2/\pi \tan^{-1}(\frac{1}{7})$.)

3.6 SOME APPLICATIONS OF HARMONIC FUNCTIONS

We shall illustrate the application of harmonic functions by some examples in hydrodynamic theory, and we begin this section by outlining briefly the connection between these two subjects. The same examples could equally well be used in the context of electrical or gravitational potential.

We suppose initially that we are dealing with a two-dimensional motion of a fluid in a region R: in other words, that at each point of R, there are given two functions u and v which are the components of velocity of the fluid at x, y. We assume also that u and v, though dependent on position in R, do not vary in time, so that we are dealing with a steady flow.

There are two kinds of conditions which u and v must satisfy. The first kind hold throughout R and arise because of the nature of the fluid and its movement, while the second concerns the behaviour of the fluid at the boundary of R and reflect the existence of solid obstacles to the flow or in the case of an unbounded region, the supposed behaviour 'at infinity'.

We begin with the conditions which hold in the interior of R.

Consider a small square subregion $S = ABCD$ of the region R whose sides are of length h and are parallel to the co-ordinate axes as in Fig. 3.5. If we let $A = (x_0, y_0)$, the fluid flowing across AB in time t is given approximately by $v(x_0, y_0)ht$, and that across DC is $v(x_0, y_0 + h)ht$. Then the net outflow in this

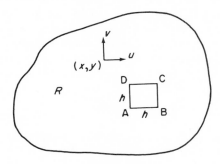

Fig. 3.5.

direction is $(v(x_0, y_0 + h) - v(x_0, y_0))ht$, which for small h is approximately

$$h\frac{\partial v}{\partial y}ht.$$

Similarly, the net outflow across AD and BC is $(u(x_0 + h, y_0) - u(x_0, y_0))ht$, which is approximately

$$h\frac{\partial u}{\partial x}ht.$$

We now make the physical assumption that the fluid is incompressible, so that the net outflow over the whole of S is zero: on allowing h to tend to zero, we obtain our first condition, namely

$$\frac{\partial u}{\partial x} + \frac{\partial v}{\partial y} = 0. \tag{3.4}$$

A second condition can be derived similarly from the assumption that the flow around ABCD is zero, which corresponds in physical terms to an absence of rotation or vorticity in the fluid. At the end of this chapter we show how this restriction can be removed. We measure the flow around ABCD by the integral $\int_{ABCD}(u\,dx + v\,dy)$ which for small h is approximately

$$hu(x_0 + \tfrac{1}{2}h, y_0) + hv(x_0 + h, y_0 + \tfrac{1}{2}h) - hu(x_0 + \tfrac{1}{2}h, y_0 + h) - hv(x_0, y_0 + \tfrac{1}{2}h),$$

where we have replaced u on AB for instance by its (constant) value $u(x_0 + \tfrac{1}{2}h, y_0)$ at the mid-point of AB. This expression is approximately

$$h\left[-h\frac{\partial u}{\partial y} + h\frac{\partial v}{\partial x} \right]$$

so that on allowing h to tend to zero we obtain our second condition.

$$-\frac{\partial u}{\partial y} + \frac{\partial v}{\partial x} = 0. \tag{3.5}$$

(Readers with a knowledge of vector calculus will recognize (3.4) and (3.5) as stating that Div $V = 0$, Curl $V = 0$, respectively where $V = (u, v)$.)

If we combine (3.4) and (3.5) we see that u and v are harmonic functions and (by comparing (3.4) and (3.5) with the Cauchy–Riemann equations (3.3) of Section 3.1) that $-v$ is the harmonic conjugate of u. It follows from Theorem 3.16 that an analytic function f exists with $f' = u - iv$. For our present purposes it is convenient to write $\phi = -\operatorname{Re} f$, thus obtaining a harmonic function with

$$u = -\frac{\partial \phi}{\partial x}, v = -\frac{\partial \phi}{\partial y}$$

(or $(u, v) = -\operatorname{Grad} \phi$ in the notation of vector calculus).

The function ϕ is called a *velocity potential* and its existence allows us to determine the flow by means of a single scalar-valued function, as will be illustrated in the examples below. The introduction of such a function is due to Euler: the minus sign (in $\phi = -\operatorname{Re} f$, etc.) is a matter of convention and means that the direction of flow is from larger to smaller values of ϕ in accordance with the usual convention of potential theory.

The curves $u = \text{constant}$ are called *equipotentials* and their orthogonal trajectories $v = \text{constant}$ give the direction of flow at any point and are called *streamlines* for the flow.

We turn now to the conditions which apply on the boundary of R. We shall suppose that the boundary of R is made up of a finite number of smooth curves (curves with a continuously varying tangent) which join at a number of vertices where discontinuities of the derivative may occur. Since we assume the boundary of R to be impenetrable, this means that at the boundary of R the direction of the flow must be parallel to the boundary, or equivalently that the normal component of the velocity must be zero. Since the velocity in any direction is given by the (negative of the) derivative of ϕ in that direction, this condition may be written

$$\frac{\partial \phi}{\partial n} = 0 \tag{3.6}$$

where n denotes a direction perpendicular to the boundary of R: see the examples following for the use of this condition. When R is unbounded we shall assume that the velocities remain bounded and will usually approach a constant value (which may be zero) as x, y become large.

One final observation is appropriate here before we begin to look at specific examples. We saw in the preceding section (and in Theorem 3.19 in particular) that if f is a one-to-one mapping of a region R_1 onto a region R_2 then ϕ is harmonic on R_2 if and only if $\phi \circ f$ is analytic on R_1. Hence the knowledge of the solution to a fluid-flow problem on a particular region R_1 can be converted into the solution of an analogous problem on R_2: our knowledge of analytic mappings from Examples 3.21 and 3.22 will thus be of great value.

We begin with a trivial example, which by an application of this principle enables us to deduce several other non-trivial cases.

Example 3.24 (Uniform flow in a plane, or half-plane). Consider the example where $\phi(x, y) = cx$ for a positive constant c.

We have
$$u = -\frac{\partial \phi}{\partial x} = -c, \qquad v = -\frac{\partial \phi}{\partial y} = 0,$$

so that this velocity potential represents a uniform flow with velocity c in the direction of the negative x-axis. The equipotentials are the lines $x = \text{constant}$, and the streamlines are $y = \text{constant}$.

This flow may occupy the whole (x, y) plane, or any half-plane of the form $\{(x, y): y > a\}$ or $\{(x, y): y < a\}$ for any real a: in particular, $a = 0$ gives the upper and lower half-planes in $\mathbb{R}^2$ respectively.

Example 3.25 (Flow in a sector) Suppose we have to find the flow in a sector of

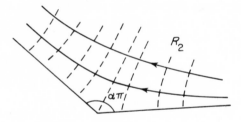

Fig. 3.6.

opening $\alpha\pi$ (Fig. 3.6). We shall assume that $0 < \alpha < 1$ so that the sector is at least convex, since the physical model we have been considering breaks down in the case of a reflex angle: for details the reader may consult Bateman (1932).

The mapping $f(z) = z^\alpha$ maps $R_1 = \{z: \operatorname{Im} z > 0\}$ onto R_2 since a point z in R_1 may be written $z = re^{i\theta}$ for $0 < \theta < \pi$, and hence $z^\alpha = r^\alpha e^{i\alpha\theta}$ is in R_2. The required flow now follows from the preceding example, where the velocity potential $\phi(z) = cx = c \operatorname{Re} z$ gives the uniform flow on R which is bounded at infinity. It follows that the velocity potential ψ in R_2 is given by $\psi = \phi(f^{-1})$, or on putting $w = \rho e^{i\beta}$ ($0 < \beta < \alpha\pi$) for a general point of R_2, that

$$\psi(w) = c \operatorname{Re}(w^{1/\alpha}) = c \operatorname{Re}(\rho^{1/\alpha} e^{i\beta/\alpha}) = c\rho^{1/\alpha} \cos(\beta/\alpha).$$

For instance, when $\alpha = \frac{1}{2}$ and the sector is a right angle we have

$$\psi(w) = cr^2 \cos 2\beta,$$

or if we put $w = a + ib$, then

$$\psi(w) = cr^2(\cos^2\beta - \sin^2\beta)$$
$$= c(a^2 - b^2).$$

It follows that the equipotential curves are the rectangular hyperbolas $a^2 - b^2 = \text{constant}$, while the streamlines are their orthogonal trajectories (also rectangular hyperbolas) given by $ab = \text{constant}$.

A further example derived from Example 3.24 is that of uniform flow around a circular obstacle. In three dimensions this corresponds to the flow of fluid past a cylindrical rod: in exercise 18 at the end of the chapter we also determine the three-dimensional flow past a sphere.

Example 3.26 (Flow past a circular obstacle) We require a harmonic function which has both the real axis and the unit circle as streamlines (Fig. 3.7). Since the flow is plainly symmetric we need only consider the region $R = \{z: \operatorname{Im} z > 0, |z| > 1\}$, and find a mapping of R onto the upper half-plane. The mapping can be found using the result of Example 3.23, as indicated in Fig. 3.8.

The result is that the mapping is given by

$$f(z) = i\left(\frac{1 - t}{1 + t}\right),$$

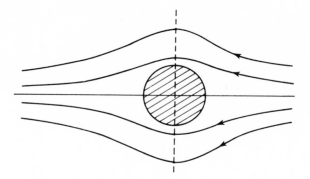

Fig. 3.7.

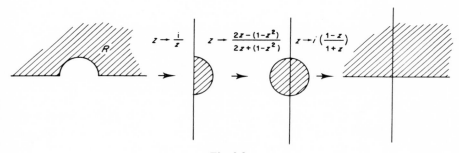

Fig. 3.8

where

$$t = \frac{2(i/z) - (1 + 1/z^2)}{2(i/z) + (1 + 1/z^2)}$$

and this simplifies to give

$$f(z) = i\frac{1 + 1/z^2}{(2i/z)} = \tfrac{1}{2}(z + 1/z).$$

(This simple answer can also be verified directly.) Consequently, the velocity potential of R is given by

$$\phi(z) = c\,\mathrm{Re}\left(z + \frac{1}{z}\right) = c\left(x + \frac{x}{x^2 + y^2}\right)$$
$$= c\left(r + \frac{1}{r}\right)\cos\theta$$

on putting $z = re^{i\theta}$, c being as usual the constant flow at infinity.

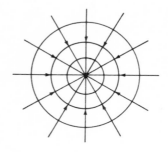

Fig. 3.9.

The streamlines are given by $\mathrm{Im}\,(z + 1/z) = \text{constant}$, or $(r - 1/r)\sin\theta = \text{constant}$, and this is satisfied by $\theta = 0, \pi$ or $r = 1$, as required.

We finish this chapter by describing a situation in which one or other of the conditions which we imposed on the fluid flow at the beginning of this section no longer hold. We saw in Example 3.18 that the function

$$\phi(x, y) = \log r = \tfrac{1}{2}\log(x^2 + y^2)$$

is harmonic in the region R obtained by deleting the origin from $\mathbb{R}^2$. If we regard ϕ as a velocity potential then the equipotential curves are circles centred at the origin, and the streamlines are straight lines converging on the origin (Fig. 3.9).

Such a flow of fluid is called a *sink* and corresponds to a point where fluid is removed from the system. The strength of a sink is measured by the line integral of the velocity crossing a circle around the sink. In this example we have, on the circle $x^2 + y^2 = r^2$, that the velocity is

$$\frac{\partial \phi}{\partial r} = \frac{1}{r}$$

in an inward direction, and hence the integral around the circle is

$$\frac{1}{r}2\pi r = 2\pi$$

(notice that the result is independent of the radius of the circle). Hence the velocity potential $\phi = \log r$ gives a sink of strength 2π, and for this reason it is convenient to define a 'unit sink' as one with velocity potential $(1/2\pi)\log r$. By changing the sign of ϕ we get a situation in which fluid flows outward from the origin, corresponding to a *source* of incoming fluid. If we refer to the discussion at the beginning of this section we see that the condition that the integral around any closed curve not passing through the origin should be zero is still satisfied, but that the net outflow across a circle which contains the source or sink is not zero.

Example 3.18 also showed that $\phi(x, y) = \tfrac{1}{2}\log(x^2 + y^2)$ has no harmonic conjugate in R. However, if we delete from R any half-line away from the origin we obtain a conjugate in this region: for instance, if we delete the negative real axis then the conjugate function is simply Arg z. This gives a flow in which the concentric circles around the origin are now the streamlines, so that the flow rotates around the origin at which point three is a *vortex*. In this case the line integral of the velocity around a circle which contains the vortex in its interior is non-zero, while the outflow across any closed curve (not through the origin) remains zero.

We can build up more complicated examples of flow in which sinks, vortices, and any other types of flow considered in this section are all present together, simply by adding together the corresponding velocity potentials, as in exercise 19.

EXERCISES

1 Let f be a continuous function on $\mathbb{R}$ for which

$$f(x) = \tfrac{1}{2}[f(x+h) + f(x-h)] \qquad \text{for all real } x, \text{ and } h > 0.$$

Show that f is given by $f(x) = a + bx$ for some real constants a and b.
2 Show that the equation of a curve in the complex plane represents a circle or a straight line if and only if it can be written in the form $Az\bar{z} + Bz + \bar{B}\bar{z} + C = 0$, for real A, C.
3 Let $f_1(e^{i\theta})$ be defined on the unit circle by

$$f_1(e^{i\theta}) = \begin{cases} 1 & \text{for } -\alpha < \theta < \alpha, \text{ where } 0 < \alpha < \pi. \\ 0 & \text{elsewhere on } [-\pi, \pi] \end{cases}$$

Find the Poisson integral of f_1 as in Example 3.8, and investigate its limiting values as $r \to 1_-$.
4 Let f_1 have right- and left-hand limits l_1 and l_2 at θ_0. Show that $f(re^{i\theta_0}) \to \tfrac{1}{2}(l_1 + l_2)$ as $r \to 1_-$. More generally, show that $f(re^{i\theta_0}) \to L$ as $r \to 1_-$ if $\tfrac{1}{2}\{f_1(e^{i(\theta_0+h)}) + f_1(e^{i(\theta_0-h)})\} \to L$ as $h \to 0$.
Hint: modify the proof of Theorem 3.6 along the lines of Theorem 2.3(ii) and (iii)).
5 Find whether each of the following series is P-summable, and if so, find its sum.

(i) $\displaystyle\sum_{n=0}^{\infty} (-1)^n(n+1)^2 = 1 - 4 + 9 - 16 + 25 - \cdots$

(ii) $1 + 1 + 1 + \cdots + 1 - 1 - 1 - 1 - \cdots - 1 + 1 + \cdots$, where r consecutive terms of $+1$ alternate with the same number of terms of -1.
6 Show that the following functions are harmonic in the regions indicated, either by verifying that they satisfy Laplace's equation or by finding an analytic function f with $u = \text{Re} f$. In each case find a harmonic conjugate v.
(i) $R = \mathbb{R}^2 : u(x,y) = x^2y - \tfrac{1}{3}y^3 + y$
(ii) $R = \mathbb{R}^2, u(x,y) = \cos x \cosh y$
(iii) $R = \{x + iy: x > 0\}, u(x,y) = \tan^{-1}(y/x), -(\pi/2) < u < (\pi/2)$.
7 Obtain the formula for the construction of v in Theorem 3.16 as follows. Suppose that a function $f = u + iv$ exists: then $f' = u_x + iv_x = u_x - iu_y$ by the

Cauchy–Riemann equations. Then $f(z) = \int_{z_0}^{z} f'(w)\,dw$; in this equation take a path from $z_0 = (x_0, y_0)$ to $z = (x, y)$ of the type considered in the proof of Theorem 3.16, when $dw = dx$ or $i\,dy$ respectively. Hence obtain the formula for v on equating imaginary parts.

8 Sum the series indicated in Example 3.17(iv) to obtain
$$Q_r(\theta) = \frac{2r\sin\theta}{1 - 2r\cos\theta + r^2}.$$

9 (Weierstrass approximation theorem) Given a continuous function f on $[0, 2\pi]$, and $\varepsilon > 0$, show that there is a trigonometric polynomial T with $|f(x) - T(x)| < \varepsilon$ for all x in $[0, 2\pi]$.
(*Hint*: find $r < 1$ so that $f(x) - f * P_r(x)$ is uniformly small and then take as T a partial sum $\sum_{n=-N}^{N} r^{|n|} c_n e^{inx}$ of $f * P_r$.) Deduce that for any continuous g on an interval $[a, b]$ and $\varepsilon > 0$, there is an (algebraic) polynomial $P(x) = \sum_{0}^{N} a_n x^n$ with $|g(x) - P(x)| < \varepsilon$ for all x in $[a, b]$.
(*Hint*: expand the terms in $T(x)$ – exponential or trigonometric – into power series.)

10 Let f be monotone on the subinterval (a, b) or $[0, 2\pi]$, and zero elsewhere. Show that there is a number K with $|c_n| < K/|n|$ for all $n \neq 0$.
(*Hint*: Lemma 2.10(ii).) Deduce the corresponding inequality for the Fourier coefficients of any piecewise monotone function.

11 Let f be an FC-function on $[0, 2\pi]$ for which $|c_n| < K/|n|$ for all $n \neq 0$. (Exercise 10 gives a wide class of functions with this property.)
The following sequence of results shows that the partial sums of $S(f)$ are uniformly bounded: compare the special method used in Lemma 2.8(i)
(i) Show that for any $r < 1$,
$$|f_r(x)| = |f * P_r(x)| \leqslant M = \sup\{|f(x)| : 0 \leqslant x \leqslant 2\pi\}.$$
(ii) Write $f_r(x) - S_n(x) = -\sum_{m=-n}^{n} c_m(1 - r^{|m|})e^{imx} + \sum_{m>|n|} c_m r^{|m|} e^{imx}$
$$= S_1 + S_2 \text{ say.}$$

Estimate S_1 using $(1 - r^n) = (1 - r)(1 + r + \cdots + r^{n-1}) < n(1 - r)$ and $|mc_m| \leqslant K$ to obtain $|S_1| \leqslant 2n(1 - r)K$.
Estimate S_2 using $|mc_m| \leqslant K$ again to obtain
$$|S_2| \leqslant \frac{2K}{n}\sum_{n+1}^{\infty} r^m < \frac{2K}{n(1 - r)}.$$

Hence choosing $1 - r = 1/n$ shows that both S_1 and S_2 are bounded by $2K$, and the result follows.

12 Let $u(x, y)$ be defined and twice continuously differentiable in the unit disc, and let $u(x, y) = U(r, \theta)$, where $x = r\cos\theta$, $y = r\sin\theta$, $1 \geqslant r \geqslant 0$, $0 \leqslant \theta \leqslant 2\pi$. Show successively that
$$\frac{\partial U}{\partial r} = \frac{\partial u}{\partial x}\cos\theta + \frac{\partial u}{\partial y}\sin\theta, \qquad \frac{\partial U}{\partial \theta} = -\frac{\partial u}{\partial x}r\sin\theta + \frac{\partial u}{\partial y}r\cos\theta,$$

and that

$$\frac{\partial^2 U}{\partial r^2} = \frac{\partial^2 u}{\partial x^2}\cos^2\theta + 2\frac{\partial^2 u}{\partial x \partial y}\sin\theta\cos\theta + \frac{\partial^2 u}{\partial y^2}\sin^2\theta,$$

$$\frac{\partial^2 U}{\partial \theta^2} = -\frac{\partial u}{\partial x}r\cos\theta - \frac{\partial u}{\partial y}r\sin\theta + \frac{\partial^2 u}{\partial x^2}r^2\sin^2\theta$$

$$-2\frac{\partial^2 u}{\partial x \partial y}r^2\sin\theta\cos\theta + \frac{\partial^2 u}{\partial y^2}r^2\cos^2\theta.$$

Deduce that the Laplace operator $\partial^2 u/\partial x^2 + \partial^2 u/\partial y^2$ is given in polar co-ordinates by

$$\frac{\partial^2 U}{\partial r^2} + \frac{1}{r}\frac{\partial U}{\partial r} + \frac{1}{r^2}\frac{\partial^2 U}{\partial \theta^2} = \frac{1}{r}\frac{\partial}{\partial r}\left(r\frac{\partial U}{\partial r}\right) + \frac{1}{r^2}\frac{\partial^2 U}{\partial \theta^2}.$$

13 Let u be harmonic in a region G and let a be a point of G. Show that when the disc centred at a radius R is contained in G, then

$$\phi(r) = \frac{1}{2\pi}\int_0^{2\pi} u(a + re^{i\theta})\,d\theta = u(a) \text{ for } 0 < r < R.$$

(*Hint*: Consider

$$r\frac{d}{dr}\left(r\frac{d\phi(r)}{dr}\right).$$

Using the result on differentiation under the integral sign (Theorem A32) and the result of exercise 12 above, this can be written as

$$-\frac{1}{2\pi}\int_0^{2\pi}\frac{\partial^2 u}{\partial \theta^2}(a + re^{i\theta})\,d\theta = -\frac{1}{2\pi}\frac{\partial u}{\partial \theta}(a + re^{i\theta})\Big|_0^{2\pi} = 0.$$

Hence $r\,d\phi(r)/dr$ must be a constant, which turns out to be zero when we let $r \to 0$. Hence also $\phi(r)$ is constant, and its value is $u(a)$, again letting $r \to 0$.)
(Notice that this exercise supplies one half of the equivalence of the mean value property of harmonic functions with their definition in terms of the Laplace operator, which we mentioned in Section 3.1.)

14 Let u be real valued and harmonic on a region G and let a be any point of G. Show that u cannot have a strict local maximum at a, and can only have a local maximum at a if u is constant.
(*Hint*: apply exercise 13 above in a suitable neighbourhood of a.) Deduce that if G contains a closed disc D, then u must attain its maximum (relative to D) at a point on the boundary of D. (This result is called the *local maximum property* of a harmonic function.)

15 Let u_1, u_2 be harmonic functions which have the same values at all points of the boundary of a closed disc D. Apply exercise 14 above to (the real and imaginary parts of) $u_1 - u_2$ to deduce that $u_1 = u_2$ in the interior of D.

16 Let f be harmonic on the interior of an equilateral triangle, and have

boundary values equal to zero on two of the sides and unity on the third side. Use Example 3.22(iii) to find the value of f at the centre of the triangle.

17 Determine the equipotentials and streamlines for a flow in sectors of opening $\frac{1}{3}\pi$, $\frac{1}{4}\pi$.

18 In three dimensions, spherical polar co-ordinates are given by

$$x = r\sin\theta\cos\phi,$$

$$y = r\sin\theta\sin\phi,$$

$$z = r\cos\theta$$

for $r \geqslant 0$, $0 \leqslant \theta \leqslant \pi$, $0 \leqslant \phi \leqslant 2\pi$. For a function $u(x, y, z) = U(r, \theta, \phi)$, the Laplace operator

$$\frac{\partial^2 u}{\partial x^2} + \frac{\partial^2 u}{\partial y^2} + \frac{\partial^2 u}{\partial z^2}$$

is given by

$$\frac{1}{r^2}\frac{\partial}{\partial r}\left(r^2\frac{\partial U}{\partial r}\right) + \frac{1}{r^2\sin\theta}\frac{\partial}{\partial\theta}\left(\sin\theta\frac{\partial U}{\partial\theta}\right) + \frac{1}{r^2\sin^2\theta}\frac{\partial U}{\partial\phi^2}.$$

(This may be shown by a calculation analogous to that in exercise 12, or may be taken on trust !).

Find the values of n for which $U = r^n\cos\theta$ gives a solution which is independent of ϕ. Hence show that the flow around a spherical obstacle (or alternatively the flowlines caused when a spherical object moves in a straight line through a stationary fluid) is given by the velocity potential

$$U = c\left(r - \frac{a^3}{2r^2}\right)\cos\theta,$$

where a is the radius of the sphere and c the velocity of the fluid when r is large.

19 (i) Find the flow in the half-plane $\{(x, y): y > 0\}$ determined by a source of strength 1 at the point $(0, 1)$.
(*Hint*: consider the flow in $\mathbb{R}^2$ resulting from two equal sources at $(0, 1)$ and $(0, -1)$.)

(ii) Deduce from (i) the flow in the quadrant $\{(x, y): x > 0, y > 0\}$ determined by a source of strength 1 at the point $(1, 1)$.

CHAPTER 4

Conjugate Functions and Conjugate Series

4.1 CONJUGATE SERIES

In Chapters 1 and 2 we investigated in some detail the relationship between a function and its Fourier Series which we denoted symbolically by

$$f(x) \sim Sf(x) = \frac{1}{2}a_0 + \sum_{n=1}^{\infty} (a_n \cos nx + b_n \sin nx),$$

with coefficients (a_n), (b_n) given by the Fourier formulae (Definition 1.5).

We now associate with this series, another which we shall denote $\tilde{S}f$.

Definition 4.1   Let f be an FC-function on $[0, 2\pi]$ and let $(a_n)_0^\infty$, $(b_n)_0^\infty$, be its Fourier coefficients.

The series

$$\sum_{n=1}^{\infty} (a_n \sin nx - b_n \cos nx)$$

is called the *conjugate Fourier series* of f and is denoted by $\tilde{S}f(x)$.

The motivation for this apparently arbitrary definition can be seen as follows. For a real-valued function f the coefficients are real and we have

$$a_n \cos nx + b_n \sin nx = \text{Re}\,(a_n - ib_n)(\cos nx + i \sin nx)$$
$$= \text{Re}\,(2c_n e^{inx}) \qquad \text{for } n \geqslant 1.$$

Hence the terms $a_n \sin nx - b_n \cos nx$, which are equal to the imaginary part of $(a_n - ib_n)(\cos nx + i \sin nx)$, form a series $\tilde{S}f(x)$ for which

$$Sf(x) + i\tilde{S}f(x) = \tfrac{1}{2}a_0 + \sum_{1}^{\infty}(a_n - ib_n)e^{inx}$$

$$= c_o + 2\sum_{1}^{\infty} c_n e^{inx}$$

89

has the form of a complex Taylor series in $z = e^{inx}$. (For the expression of $\tilde{S}f$ directly in terms of the complex Fourier series of f, see exercise 1 at the end of the chapter.)

A second reason for this definition comes from the discussion of harmonic conjugates in Chapter 3. In Section 3.4 we found that if f is defined on the boundary of the unit disc then its Poisson integral $u(re^{i\theta})$ is a harmonic function given by $u(re^{i\theta}) = f * P_r(\theta)$, and the harmonic conjugate $v(re^{i\theta})$ of u is given by $v(re^{i\theta}) = f * Q_r(\theta)$, where

$$P_r(\theta) = \frac{1 - r^2}{1 - 2r\cos\theta + r^2}, \qquad Q_r(\theta) = \frac{2r\sin\theta}{1 - 2r\cos\theta + r^2}$$

are the Poisson kernel and conjugate Poisson kernel respectively.

We saw in Chapter 3 that

$$P_r(\theta) = \sum_{-\infty}^{\infty} r^{|n|} e^{in\theta} = 1 + 2\sum_{1}^{\infty} r^n \cos n\theta$$

$$= \mathrm{Re}\left(\frac{1 + z}{1 - z}\right) \qquad (z = re^{i\theta})$$

and that

$$Q_r(\theta) = -i\sum_{1}^{\infty} r^n(e^{in\theta} - e^{-in\theta})$$

$$= 2\sum_{1}^{\infty} r^n \sin n\theta = \mathrm{Im}\left(\frac{1 + z}{1 - z}\right).$$

It follows that

$$u(re^{i\theta}) = \sum_{-\infty}^{\infty} r^{|n|} c_n e^{in\theta}$$

and that

$$v(re^{i\theta}) = -i\sum_{1}^{\infty} r^n(c_n e^{in\theta} - c_{-n} e^{-in\theta}) \qquad \text{(recall Lemma 1.18(iii)).}$$

Hence $F(re^{i\theta}) = (u + iv)(re^{i\theta}) = c_0 + 2\sum_{1}^{\infty} c_n r^n e^{in\theta}$ is an analytic function of $z = re^{i\theta}$, and when f is real, u and v are its real and imaginary parts. On putting $r = 1$ we obtain formally

$$F(e^{i\theta}) = c_0 + 2\sum_{1}^{\infty} c_n e^{in\theta} = S(f) + i\tilde{S}(f) \qquad \text{as before.}$$

This suggests that we might consider a conjugate *function* $\tilde{f}$ which could be defined in one of two ways: either as the sum of $\tilde{S}f$, should the series converge, or as the result of putting $r = 1$ in the above formula for $v(re^{i\theta})$. Since

$$Q_1(\theta) = \frac{2\sin\theta}{2 - 2\cos\theta} = \frac{2\sin\frac{1}{2}\theta\cos\frac{1}{2}\theta}{2\sin^2\frac{1}{2}\theta} = \cot\frac{1}{2}\theta,$$

this gives us

$$\tilde{f}(\theta) = \frac{1}{2\pi} \int_0^{2\pi} f(t) \cot \tfrac{1}{2}(\theta - t) \, dt,$$

an integral which may be written in the alternative forms

$$\frac{1}{2\pi} \int_0^{2\pi} f(\theta - t) \cot \tfrac{1}{2} t \, dt = \frac{1}{2\pi} \int_0^{\pi} \{f(\theta - t) - f(\theta + t)\} \cot \tfrac{1}{2} t \, dt$$

on using the periodicity of f and the fact that cost is an odd function.

Since $\cot \tfrac{1}{2} t$ is not integrable, some caution is needed in using this integral as a definition; hence the need for the more cautious approach which we shall adopt from now on.

Definition 4.2 Let f be an FC-function on $[0, 2\pi]$. We define $\tilde{f}(x)$ to be the limit as $\delta \to 0$ (if it exists) of the integral

$$\frac{1}{2\pi} \int_\delta^{\pi} \{f(x - t) - f(x + t)\} \cot \tfrac{1}{2} t \, dt.$$

We can now pose the following questions, to which the remainder of the chapter will be devoted:

1. Under what conditions of f and x does the integral in Definition 4.2 have a limit?
2. Similarly, for what f and x is the series $\tilde{S}(f)(x)$ convergent, or P-summable?
3. Assuming the existence of the limits in 1 and 2 above, when do the values agree?
4. What properties of $\tilde{f}$ (e.g. continuity) can be deduced from those of f?

Before tackling these problems, we consider a few examples.

Example 4.3 (i) Let h be a fixed real number between 0 and π, and let

$$f(x) = \begin{cases} 1 & \text{for } |x| \leqslant h, \\ 0 & \text{for } h < |x| \leqslant \pi. \end{cases}$$

Then it is easy to find

$$S(f)(x) = \frac{h}{\pi} + \frac{2}{\pi} \sum_1^{\infty} \frac{\sin(nh)}{n} \cos nx,$$

and it follows from Definition 4.1 that

$$\tilde{S}(f)(x) = \frac{2}{\pi} \sum_1^{\infty} \frac{\sin(nh)}{n} \sin nx.$$

If we use the elementary expansion $\log(1 - z) = -\sum_1^{\infty} (1/n) z^n$ and put

$$z = r e^{i\theta}, \qquad 0 \leqslant r < 1,$$

we obtain

$$\log|1 - re^{i\theta}| + i\,\text{Arg}(1 - re^{i\theta}) = -\sum_{1}^{\infty}\frac{1}{n}r^n(\cos n\theta + i\sin n\theta).$$

In particular if as in Chapter 3, we let $r \to 1$, write

$$1 - e^{i\theta} = 2\sin\tfrac{1}{2}\theta e^{i(\theta - \pi)/2},$$

and equate real and imaginary parts, we get

$$\log 2|\sin\tfrac{1}{2}\theta| = -\sum_{1}^{\infty}\frac{1}{n}\cos n\theta \text{ and } \tfrac{1}{2}(\pi - \theta) = \sum_{1}^{\infty}\frac{1}{n}\sin n\theta, \qquad 0 < \theta < 2\pi.$$

The second of these results is of course well known from Chapter 1, while the first is new (notice that $\log 2|\sin\tfrac{1}{2}\theta|$ is not an FC-function and so does not come strictly within the scope of Chapters 1 and 2).

If we write the series

$$\tilde{S}(f) = \frac{2}{\pi}\sum_{1}^{\infty}\frac{\sin nh}{n}\sin nx$$

in the form

$$\frac{1}{\pi}\sum_{1}^{\infty}\frac{1}{n}(\cos n(x - h) - \cos n(x + h))$$

we see that it is convergent except where $x = \pm h$, and has sum

$$\frac{1}{\pi}\log\left|\frac{\sin\tfrac{1}{2}(x + h)}{\sin\tfrac{1}{2}(x - h)}\right|.$$

This expression is also the value of the integral

$$\frac{1}{\pi}\int_{0}^{\pi}(f(x - t) - f(x + t))\cot\tfrac{1}{2}t\,dt$$

(exercise 3 at the end of the chapter) and is thus the value of $\tilde{f}(x)$ for all x except $\pm h$. If we compare this result with the questions 1–4 above we see that the series $\tilde{S}(f)$ and the integral for $\tilde{f}$ converge and have the same values for the same points of the interval. Notice, however, that f is an FC-function, having discontinuities only at $\pm h$, while $\tilde{f}$ is unbounded in the neighbourhood of these points.

(ii) Let $f(x) = \dfrac{1}{a - \cos x}, \quad a > 1$

If we choose $r = a - \sqrt{a^2 - 1}$ so that $0 < r < 1$ and $1 + r^2 = 2ar$ then

$$f(x) = \frac{2r}{1 + r^2 - 2r\cos x} = \frac{2r}{1 - r^2}P_r(x) = \frac{2r}{1 - r^2}\left(1 + 2\sum_{1}^{\infty}r^n\cos nx\right).$$

The conjugate function is then found from

$$\frac{2r}{1-r^2}Q_r(x)=\frac{2r}{1-r^2}2\sum_1^\infty r^n\sin nx$$

$$=\frac{2r}{1-r^2}\frac{2r\sin x}{1+r^2-2\cos rx}=\frac{2r}{1-r^2}\frac{\sin x}{a-\cos x}$$

$$=(a^2-1)^{-\frac12}\frac{\sin x}{a-\cos x}.$$

Notice that in this case, as in most others in practice, it is easier to find the conjugate series and sum it, then to evaluate the defining integral in Definition 4.1 directly. (In so doing, we are anticipating the result of Theorem 4.5 below.)

When we come to investigate the questions which we raised concerning the existence of $\tilde{f}(x)$ or the convergence of $\tilde{S}(f)$, we find that everything goes well in the presence of Lipschitz continuity, but that once this is dropped, many difficulties arise. We begin with a lemma which is analogous to Lemma 2.1.

Lemma 4.4 Let $\tilde{D}_n(x)=2\sum_{r=1}^n\sin rx$.
Then

$$\tilde{D}_n(x)=\begin{cases}0 & \text{if } x \text{ is a multiple of } 2\pi,\\(\sin\tfrac12 x)^{-1}\{\cos\tfrac12 x-\cos(n+\tfrac12)x\} & \text{if not.}\end{cases}$$

Also the partial sum

$$\tilde{S}_n(f,x)=\sum_{r=1}^n(a_r\sin rx-b_r\cos rx)$$

of $\tilde{S}(f)$ can be written as

$$f*\tilde{D}_n(x)=\frac{1}{2\pi}\int_0^{2\pi}f(t)\tilde{D}_n(x-t)\,\mathrm{d}t=\frac{1}{2\pi}\int_0^\pi\tilde{D}_n(t)\{f(x-t)-f(x+t)\}\,\mathrm{d}t$$

Proof If x is not a multiple of 2π, $\sin\tfrac12 x$ is non-zero and we may write

$$2\sum_{r=1}^n\sin rx=\frac{1}{i}\sum_1^n(e^{irx}-e^{-irx})$$

$$=\frac{1}{i}\left[\frac{e^{ix}-e^{i(n+1)x}}{1-e^{ix}}-\frac{e^{-ix}-e^{-i(n+1)x}}{1-e^{-ix}}\right]$$

$$=\frac{\sin\tfrac12 nx}{\sin\tfrac12 x}\cdot\frac{1}{i}\{e^{-i\frac12(n+1)x}-e^{-i\frac12(n+1)x}\}$$

$$=\frac{2\sin\tfrac12 nx\sin\tfrac12(n+1)x}{\sin\tfrac12 x}$$

$$=\frac{\cos\tfrac12 x-\cos(n+\tfrac12)x}{\sin\tfrac12 x}$$

which gives the required expression for $\tilde{D}_n$.

We then find that

$$\tilde{S}_n(f, x) = \sum_1^n (a_r \sin rx - b_r \cos rx)$$

$$= \frac{1}{2\pi} \sum_1^n \int_0^{2\pi} f(t)(\cos rt \sin rx - \sin rt \cos rx) \, dt$$

$$= \frac{1}{2\pi} \int_0^{2\pi} f(t) \sum_1^n \sin r(x - t) \, dt = f * \tilde{D}_n(x).$$

We can also write this as

$$\frac{1}{2\pi} \int_{-\pi}^{\pi} \tilde{D}_n(t) f(x - t) \, dt$$

which is equal to

$$\frac{1}{2\pi} \int_0^{\pi} \tilde{D}_n(t)(f(x - t) - f(x + t) \, dt \qquad \text{since } \tilde{D}_n \text{ is odd,}$$

and this completes the proof.

We now establish the main result of this section.

Theorem 4.5 Let f be an FC-function on $[0, 2\pi]$, and let x_0 be a point of the interval for which a constant M exists with

$$|f(x_0 - h) - f(x_0 + h)| \leqslant Mh \qquad \text{for all } h > 0.$$

Then (i) the conjugate series $\tilde{S}f(x)$ is convergent at x_0, with the sum s say, and (ii) the conjugate function $\tilde{f}(x)$ exists at x_0, with $\tilde{f}(x_0) = s$.

Proof Definition 4.2 says that $\tilde{f}(x_0)$ is the limit (if it exists) of the integral

$$\frac{1}{2\pi} \int_\delta^{\pi} \{f(x_0 - t) - f(x_0 + t)\} \cot \tfrac{1}{2}t \, dt,$$

as $\delta \to 0$: in particular, if the integral $\int_0^{\pi}$ exists in the usual sense (without the limit) then its value is $\tilde{f}(x_0)$. But $f(x_0 - t) - f(x_0 + t)$ is an FC-function of t, and $\cot \tfrac{1}{2}t$ is continuous on $(0, \pi]$. Hence the above integral will exist over $[0, \pi]$ provided that the integrand remains bounded near $t = 0$.

However, if we write the integrand in the form

$$\frac{f(x_0 - t) - f(x_0 + t)}{t} \cdot \frac{t}{\sin \tfrac{1}{2}t} \cdot \cos \tfrac{1}{2}t,$$

then the first factor is bounded by the hypothesis on f at x_0, while the second and third approach 2 and 1 respectively as $t \to 0$. Hence the limit which defines $\tilde{f}(x_0)$

exists and its value is

$$\frac{1}{2\pi} \int_0^\pi \{f(x_0 - t) - f(x_0 + t)\} \cot \tfrac{1}{2}t \, dt.$$

If we now consider the conjugate series $\tilde{S}f(x_0)$, Lemma 4.4 shows that

$$\tilde{S}_n(f, x_0) = \frac{1}{2\pi} \int_0^\pi \{f(x_0 - t) - f(x_0 + t)\} \tilde{D}_n(t) \, dt.$$

If we write $\tilde{D}_n(t) = (\cos \tfrac{1}{2}t - \cos(n + \tfrac{1}{2})t)/\sin \tfrac{1}{2}t$, and subtract the value of $\tilde{f}(x_0)$ which we have just found, to obtain

$$\tilde{S}_n(f, x_0) - \tilde{f}(x_0) = -\frac{1}{2\pi} \int_0^\pi \left\{ \frac{f(x_0 - t) - f(x_0 + t)}{\sin \tfrac{1}{2}t} \right\} \cos(n + \tfrac{1}{2})t \, dt.$$

However, we have just shown that

$$\frac{f(x_0 - t) - f(x_0 + t)}{\sin \tfrac{1}{2}t}$$

is an FC-function, and hence the integral on the right tends to zero as $n \to \infty$, by the Riemann–Lebesgue lemma (Corollary 1.14(iii)) Hence $\tilde{S}(f, x_0) \to \tilde{f}(x_0)$ as required.

In practice we usually find that the following special case is sufficient

Corollary 4.6 If f is Lipschitz continuous (Definition A.5) at x_0, then $\tilde{f}(x_0)$ exists, and is equal to the sum of the (convergent) series $\tilde{S}f(x_0)$.

It should be noticed that the theorem implies considerably more than the corollary: for instance if for some $h > 0$, $f(x) = \cos 1/x$ for $0 \leqslant |x| \leqslant h$, and is equal to any FC-function on the rest of $[-\pi, \pi]$, then f is not even continuous, still less Lipschitz continuous at $x_0 = 0$. However, $f(t) - f(-t) = 0$ for $0 < t < h$, so the hypothesis of the theorem is satisfied, and the existence of $\tilde{f}(0)$ and the convergence of $\tilde{S}f(0)$ are assured (though the convergence of $Sf(0)$ is certainly not!) (Exercise 6, in which the behaviour of $\tilde{S}f(x)$ near a point at which right- and left-hand Lipschitz limits exist and are distinct, should be considered here also.)

In general Theorem 4.5 serves as a reasonable guide to the existence of the conjugate function. For instance, if $f(x)$ is equal to $[\log\{x(\pi - x)\}]^{-1}$ on $(0, \pi)$ and zero on $[-\pi, 0]$, then f is continuous on $[0, 2\pi]$ but not Lipschitz continuous, and the integral defining $\tilde{f}(0)$, namely

$$\frac{1}{2\pi} \int_0^\pi [\log\{t(\pi - t)\}]^{-1} \cot \tfrac{1}{2}t \, dt$$

does not exist (even in the weaker form $\lim_{\delta \to 0} \int_\delta^\pi$) due to the divergence of the integral

$$\int_0^1 \frac{dt}{t \log t}.$$

Consequently $\tilde{f}$ does not exist at $t = 0$ for this function.

Questions about the existence of $\tilde{f}(x)$ at 'most' points of the interval $[0, 2\pi]$ are beyond the scope of our techniques. For instance, it is known that for any continuous f, or even any function of the Lebesgue class L^1, the conjugate function $\tilde{f}$ exists at almost every point (in the sense of Lebesgue measure). Moreover, if f is in the Lebesgue class L^p for $1 < p < \infty$ (this means that $|f|^p$ is integrable) then $\tilde{f}$ is also in L^p, and the mapping from f to $\tilde{f}$ is bounded, and is even an isometry if $p = 2$ (for a partial indication of this last result, see Lemma 4.8 below) the corresponding result for $p = 1$ is false, however. These results are discussed in more detail in the references in Appendix C.

4.2 PROPERTIES OF THE CONJUGATE FUNCTION

This section deals with some simple cases in which properties of $\tilde{f}$ (such as continuity) can be deduced from properties of f. The main result shows that Lipschitz continuity of f (on an interval) is sufficient for continuity of $\tilde{f}$. We could establish this by a direct but difficult argument, showing that

$$\frac{1}{2\pi} \int_0^\pi (f(x - t) - f(x + t)) \cot \tfrac{1}{2} t \, dt$$

varies continuously with x. However, it turns out to be more useful to prove somewhat more, namely that when f is Lipschitz continuous on the whole of $[0, 2\pi]$ then the conjugate *series* $\tilde{S}(f)$ is *uniformly* convergent: this has two consequences, that $\tilde{f}$ is continuous and that $\tilde{S}(f)$ is its Fourier series; symbolically $\tilde{S}(f) = S(\tilde{f})$.

Notice that continuity of f alone does not ensure even the existence of $\tilde{f}$ at all points, let alone its continuity, as the example at the end of the preceding section shows. Notice also that exercise 4 at the end of the chapter shows that f may be Lipschitz continuous while $\tilde{f}$ is not.

Theorem 4.7 Let f be Lipschitz continuous on $[0, 2\pi]$. Then the conjugate series $\tilde{S}(f)$ is uniformly convergent on $[0, 2\pi]$ with sum $\tilde{f}$. In particular, $\tilde{f}$ is continuous and $\tilde{S}(f)$ is its Fourier series: $\tilde{S}(f) = S(\tilde{f})$.

Proof From Lemma 4.4 we have

$$\tilde{S}_n(f, x) = \frac{1}{2\pi} \int_0^\pi \{f(x - t) - f(x + t)\} \tilde{D}_n(t) \, dt,$$

where $\tilde{D}_n(t) = (\sin \tfrac{1}{2} t)^{-1} \{\cos \tfrac{1}{2} t - \cos (n + \tfrac{1}{2}) t\}, \qquad t \neq 0.$

Also, by the Lipschitz continuity of f, and Corollary 4.6, the integral which defines $\tilde{f}$,

$$\tilde{f}(x) = \frac{1}{2\pi} \int_0^\pi \{f(x-t) - f(x+t)\} \cot\tfrac{1}{2}t \, dt,$$

exists at each point x of $[0, 2\pi]$ and we may subtract this value to obtain

$$\tilde{S}_n(f, x) - \tilde{f}(x) = -\frac{1}{2\pi} \int_0^\pi \frac{f(x-t) - f(x+t)}{\sin\tfrac{1}{2}t} \cos(n+\tfrac{1}{2})t \, dt,$$

and consequently we have to prove that the right-hand side tends to zero *uniformly* in x as $n \to \infty$. This argument resembles the proof of Theorem 2.5, and we outline it as follows.

Suppose that $\varepsilon > 0$ is given and that we divide the range of integration $[0, \pi]$ into $[0, \delta]$ and $[\delta, \pi]$, where δ is to be determined shortly. Then

$$\left| \frac{1}{2\pi} \int_0^\delta \frac{f(x-t) - f(x+t)}{\sin\tfrac{1}{2}t} \cos(n+\tfrac{1}{2})t \, dt \right| \leqslant \frac{1}{2\pi} \int_0^\delta \frac{2Mt}{\sin\tfrac{1}{2}t} \, dt < \frac{1}{\pi} 3M\delta,$$

where (as in Theorem 2.5) we have used the estimate $t/(\sin\tfrac{1}{2}t) < 3$ for t near 0. We now fix a value of δ so that $(3/\pi)M\delta < \varepsilon$.

The other integral,

$$\frac{1}{2\pi} \int_\delta^\pi \frac{f(x-t) - f(x+t)}{\sin\tfrac{1}{2}t} \cos(n+\tfrac{1}{2})t \, dt,$$

will tend to zero by the Riemann–Lebesgue lemma provided that we can show that the integral modulus of continuity of the function $(f(x-t) - f(x+t))g(t)$ (as a function of t) tends to zero *uniformly* in x: here we have written $g(t)$ for the function which is equal to $(\sin\tfrac{1}{2}t)^{-1}$ on $[\delta, \pi]$, and zero elsewhere.

This modulus of continuity may be written

$$\int_0^{2\pi} |\{f(x-t) - f(x+t)\}g(t) - \{f(x-t-h) - f(x+t+h)\}g(t+h)| \, dt$$

$$\leqslant \int_0^{2\pi} |\{(f(x-t) - f(x-t-h)) - (f(x+t) - f(x+t+h))\}g(t)| \, dt$$

$$+ \int_0^{2\pi} |\{f(x-t-h) - f(x+t+h)\}(g(t) - g(t+h))| \, dt$$

$$\leqslant 2M_1\omega_f(h) + 2M_2\omega_g(h),$$

where M_1, M_2 denote sup $|g|, |f|$ respectively and ω_f, ω_g are the moduli of continuity of f, g. The result now follows since $\omega_f, \omega_g \to 0$ with h. The final statements of the theorem follow at once: $\tilde{f}$ must be continuous, being the sum of a uniformly convergent series of continuous terms, and $\tilde{S}(f)$ must be its Fourier series $S(\tilde{f})$ on integrating termwise.

Note A modification of the above argument shows that if f is Lipschitz continuous on a subinterval $I = (a, b)$ of $[0, 2\pi]$ then $\tilde{f}$ exists and is the (uniform) sum of $\tilde{S}(f)$ on I. However, we cannot even assert the existence of $\tilde{f}$ outside I and so cannot consider $S(\tilde{f})$ either.

The following result is immediate as stated, but has important generalizations.

Lemma 4.8 The mapping $f \to \tilde{f}$ is isometric on trigonometric polynomials which satisfy $P(0) = 0$. More explicitly, if $P(x) = \sum_{-n}^{n} c_n e^{inx}$ where $c_0 = 0$, then

$$\frac{1}{2\pi} \int_0^{2\pi} |P(x)|^2 \, dx = \frac{1}{2\pi} \int_0^{2\pi} |\tilde{P}(x)|^2 \, dx.$$

Proof If $P(x) = \sum_{-n}^{n} c_n e^{inx}$, then exercise 1 at the end of this chapter shows that

$$\tilde{P}(x) = -i \sum_1^n (c_n e^{inx} - c_{-n} e^{-inx})$$

(and there are no convergence problems here!).

Hence both the required integrals are equal to

$$\sum_{-n}^{n} |c_n|^2, \text{ since } c_0 = 0.$$

Examples 4.9 (i) Let

$$f(x) = \begin{cases} 1 & \text{on } (0, \pi), \\ -1 & \text{on } (-\pi, 0), \\ 0 & \text{at } \pm\pi, 0. \end{cases}$$

Then as in Chapters 1 and 2 we find that

$$S(f) = \frac{4}{\pi} \sum_{n=0}^{\infty} \frac{1}{2n+1} \sin(2n+1)x,$$

and that the series is convergent to f for all x. The conjugate series is

$$\tilde{S}(f) = -\frac{4}{\pi} \sum_{n=0}^{\infty} \frac{1}{2n+1} \cos(2n+1)x$$

and we may find its sum by writing it as

$$-\frac{4}{\pi} \sum_{n=0}^{\infty} \frac{1}{2n+1} \frac{1}{2} (e^{i(2n+1)x} + e^{-i(2n+1)x})$$

$$= -\frac{4}{\pi} \frac{1}{4} \left[\log\left(\frac{1+e^{ix}}{1-e^{ix}}\right) + \log\left(\frac{1+e^{-ix}}{1-e^{-ix}}\right) \right],$$

on using the expansion

$$\frac{1}{2} \log\left(\frac{1+z}{1-z}\right) = \sum_0^{\infty} \frac{1}{2n+1} z^{2n+1}.$$

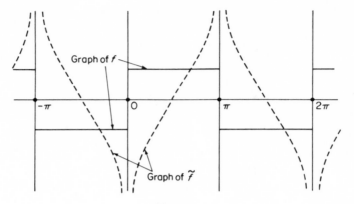

Fig. 4.1.

This simplifies to give

$$-\frac{1}{\pi}\log\left(\frac{1+\cos x}{1-\cos x}\right) = -\frac{1}{\pi}\log\frac{\cos^2\frac{1}{2}x}{\sin^2\frac{1}{2}x} = \frac{2}{\pi}\log|\tan\tfrac{1}{2}x|.$$

It follows from Corollary 4.6 that $\tilde{f}(x)$ exists and is equal to

$$\frac{2}{\pi}\log|\tan\tfrac{1}{2}x|$$

for all x not a multiple of π (see Fig. 4.1); notice that $\tilde{f}$ is continuous on $(0,\pi)$ and $(-\pi,0)$ in accordance with the note following Theorem 4.7.

(ii) Let $f(x) = \cos ax$ on $[-\pi,\pi]$, where a is any non-integral real or complex number. The Fourier series of f was found in Example 1.12(ii) where we showed that

$$\pi\frac{\cos ax}{\sin a\pi} = \frac{1}{a} + \sum_{n=1}^{\infty}\frac{(-1)^n 2a}{a^2-n^2}\cos nx.$$

The conjugate series for $\tilde{f}$ gives

$$\frac{\pi}{\sin a\pi}\tilde{S}(f)(x) = \sum_{n=1}^{\infty}(-1)^n\frac{2a}{a^2-n^2}\sin nx,$$

where the series is absolutely convergent, and its sum is continuous in accordance with Theorem 4.7.

In this example $\tilde{f}(x)$ cannot be found explicitly for general values of a, in terms of elementary functions. Exercise 4 gives a result which is equivalent to the special case $a = \frac{1}{2}$.

EXERCISES

1 Show that if f is a real- or complex-valued FC-function on $[0,2\pi]$ and

$S(f)(x) = \sum_{-\infty}^{\infty} c_n e^{inx}$, then

$$\tilde{S}(f)(x) = \frac{1}{i}\left(\sum_{1}^{\infty} c_n e^{inx} - \sum_{-\infty}^{-1} c_n e^{inx}\right)$$

$$= -i\sum_{1}^{\infty}(c_n e^{inx} - c_{-n}e^{-inx}).$$

2   Assuming that $\tilde{f}(x)$ exists for all points in $[-\pi, \pi]$, show that $\tilde{f}$ is an odd function if and only if f is even, and $\tilde{f}$ is even if and only if f is odd.

3   For $0 < h < \pi$, let $f(x) = 1$ on $[-h, h]$ and 0 elsewhere on $[-\pi, \pi]$. Show that

$$\frac{1}{\pi}\int_0^\pi \{f(x-t) - f(x+t)\}\cot\tfrac{1}{2}t\,dt$$

exists for all x except $\pm h$, and that its value is

$$\frac{1}{\pi}\log\left|\frac{\sin\frac{1}{2}(x+h)}{\sin\frac{1}{2}(x-h)}\right|$$

(*Hint*: consider $0 < x < h$, and $h < x < \pi$ separately, and use exercise 2 for negative x.) In particular, f is an FC-function, but $\tilde{f}$ is unbounded.

4   Show that if $f(x) = |\sin x|$ on $[-\pi, \pi]$, then

$$\tilde{f}(x) = \frac{2}{\pi}\sin x \log|\cot\tfrac{1}{2}x|.$$

(This example shows that f may be Lipschitz continuous, while $\tilde{f}$ is not.)
(*Hint*: use the series for $\log 2|\sin\frac{1}{2}x|$ found in Example 4.3(i).)

5   Sketch the graph of $\tilde{D}_n(x)$ and $Q_r(x)$ for $0 \leqslant x \leqslant \pi$, showing the position of any zeros and maxima or minima.
(*Hint*: $\tilde{D}_n(x) = 2\sin\frac{1}{2}nx\,\sin\frac{1}{2}(n+1)x/\sin\frac{1}{2}x$.)

6   Let f be an FC-function, and x_0 a point at which f has distinct right- and left-hand limits l_1 and l_2 say, for which $|f(x+h) - l_1| + |f(x-h) - l_2| \leqslant Mh$ for $h > 0$. Suppose also that f is Lipschitz continuous on the intervals $(x_0 - \delta, x_0)$ and $(x_0, x_0 + \delta)$ for some $\delta > 0$. Let $\phi(x)$ denote the function of Example 4.9(i). Show that $f(x) - \frac{1}{2}\Delta\phi(x - x_0)$ (where $\Delta = l_2 - l_1$) is Lipschitz continuous on $(x_0 - \delta, x_0 + \delta)$ and hence show that $\tilde{f}(x)$ is unbounded near x_0.

CHAPTER 5

The Fourier Integral

5.1 INTRODUCTION

In our earlier chapters dealing with Fourier series, we have considered a function f which represents a *periodic* phenomenon with $f(x + a) = f(x)$ (and usually $a = 2\pi$ for convenience), and tried to represent it as a sum of fundamental periodic components such as exponential or trigonometric functions. This give rise to mathematical problems concerning the series and the sense in which it represented the original function, but the underlying *form* of the representation was clear from the beginning.

This chapter begins a new area of study, namely the problem of modifying the method to deal with non-periodic phenomena. This introductory section will deal with two possible ways of discovering the form this should take, one of which is based on our existing knowledge of Fourier series, while the other uses a more abstract approach based on the idea of a group. It should be stressed that both these arguments are informal and for purposes of motivation: the systematic development will begin in Section 5.2.

Suppose then that we have some function f which is defined for all real numbers, and may be supposed to represent some non-periodic phenomenon such as the build-up and release of water in a reservoir, or the readings of a seismograph during an earthquake in which there is again a build-up followed by a period in which the readings decrease while not necessarily vanishing completely. In mathematical terms we shall suppose that $\int_{-\infty}^{\infty} |f(x)| \, dx$ has a finite positive value, i.e. that f is absolutely integrable over the real line (see also Definition A.13); we shall consider hypotheses on f more carefully in the next section.

If we recall that the integral of f, $\int_{-\infty}^{\infty} f(x) \, dx$, is by definition the limit of integrals $\int_{A}^{B} f(x) \, dx$ over subintervals of the real line, then we may hit on the idea of restricting f to a subinterval, and investigating this restricted function by means of its Fourier series, and afterwards taking some sort of limit as the size of the subinterval becomes large. With this in mind, let us suppose that our basic interval is $[-a, a]$ for some positive real number a. The exponential function

101

e^{-ixy} will have period $2a$ (in x) provided $e^{-iay} = e^{iay}$, or $2ay = 2n\pi$ for some integer n. Thus $y = n\pi/a$, and we get the system $e_n(x) = e^{-ixn\pi/a}$, $n = 0, \pm 1, \pm 2, \ldots$, corresponding to the exponential system of Lemma 1.4(i).

This means that at least for $-a \leqslant x \leqslant a$, we may expect f to be represented by its Fourier series

$$\sum_{-\infty}^{\infty} c_n(a)e^{inx\pi/a}, \tag{5.1}$$

where the coefficients are given by the Fourier formulae

$$c_n(a) = \frac{1}{2a}\int_{-a}^{a} f(x)e^{-inx\pi/a}\,dx. \tag{5.2}$$

(Notice that the factor $2a$ occurs as from the orthogonality conditions,

$$\int_{-a}^{a} e_n(x)\overline{e_m(x)}\,dx = \begin{cases} 2a & \text{if } m = n \\ 0 & \text{if } m \neq n \end{cases}$$

and the formulae of section 2, Chapter 1. The series of Chapters 1–4 are of course the special case of (5.1) and (5.2) when $a = \pi$.)

We now investigate what happens when a becomes large. If we fix n and let $a \to \infty$, the right-hand side of (5.2) is bounded by

$$\frac{1}{2a}\int_{-\infty}^{\infty} |f(x)|\,dx$$

and so tends to zero, and we can make no progress. Instead we rewrite (5.2) as

$$2ac_n(a) = \int_{-a}^{a} f(x)e^{-inx\pi/a}\,dx,$$

from which we see that if both n and a tend to infinity in such a way that

$$n/a \to 2y \qquad \text{for some real number } y, \tag{5.3}$$

then $2ac_n(a) \to \int_{-\infty}^{\infty} f(x)e^{-2\pi ixy}\,dx$, a function of y which we shall denote (temporarily) by $\phi(y)$. (The choice of the factor 2 in $(n/a) \to 2y$ is to some extent arbitrary but leads to the symmetric formulae (5.4) and (5.5) below). Thus our new version of (5.2) is the formula

$$\phi(y) = \int_{-\infty}^{\infty} f(x)e^{-2\pi ixy}\,dx \tag{5.4}$$

This integral exists for all real y since $\int_{-\infty}^{\infty} |f(x)|\,dx$ is finite.

We now turn to the new version of the series for f, (5.1) above, and write this as

$$\sum_{-\infty}^{\infty} c_n(a)e^{inx\pi/a} = \sum_{-\infty}^{\infty} \frac{1}{2a}(2ac_n(a))e^{inx\pi/a}.$$

We have to find the limit of this as both n and $a \to \infty$, while $n/a \to 2y$.

Informally, we might write $y_n = n/2a$, so that $y_{n+1} - y_n = 1/2a$, and the summation becomes approximately $\sum_{-\infty}^{\infty}(y_{n+1} - y_n)\phi(y_n)e^{2\pi ixy_n}$, and if we regard the real line as divided into strips of width $1/2a$ by the points y_n, then this summation is in turn a kind of approximation to the integral $\int_{-\infty}^{\infty} \phi(y)e^{2\pi ixy}\,dy$. Reminding ourselves (if it should be necessary) that we have as yet *proved* nothing, we see that we have arrived at the formula which gives f in terms of an integral (not a series), namely

$$f(x) = \int_{-\infty}^{\infty} \phi(y)e^{2\pi ixy}\,dy. \tag{5.5}$$

The new formulae (5.4) and (5.5) are of course only informally established, but demand attention as the appropriate analogues of the Fourier formulae in Chapter 1 (or indeed in (5.1) and (5.2) above) and by their elegance and symmetry. We shall see later in the chapter that the usual analytical problems arise – for instance, ϕ may not be absolutely integrable so that the integral on the right-hand side of (5.5) has no guaranteed existence. However, we postpone these problems for the moment and consider an example which, as well as being fundamentally important in its own right, demonstrates that the formulae (5.4) and (5.5) are correct in a non-trivial instance.

Example 5.1 Let $f(x) = e^{-\pi x^2}$. Then we shall show that $\phi(y) = e^{-\pi y^2}$ and the formulae (5.4) and (5.5) are valid.

Because of the symmetry between f and ϕ, it is sufficient to show that the integral (5.4) has the stated value, i.e. that

$$\int_{-\infty}^{\infty} e^{-\pi x^2 - 2\pi ixy}\,dx = e^{-\pi y^2}.$$

But we may write $\int_{-\infty}^{\infty} e^{-\pi x^2 - 2\pi ixy}\,dy = e^{-\pi y^2}\int_{-\infty}^{\infty} e^{-\pi(x+iy)^2}\,dx$, and so it is sufficient to show that $g(y) = \int_{-\infty}^{\infty} e^{-\pi(x+iy)^2}\,dx = 1$ for all real y.

To show this we refer to Appendix B, where it is shown (Theorem B.4) $g(0) = \int_{-\infty}^{\infty} e^{-\pi x^2}\,dx = 1$, and thus we have only to show that $g'(y) = 0$ for all y. For this, we must differentiate under the integral sign (as in Theorem 5.8 below) and we obtain

$$g'(y) = \int_{-\infty}^{\infty} -2\pi i(x+iy)e^{-\pi(x+iy)^2}\,dx$$

$$= i[e^{-\pi(x+iy)^2}]_{-\infty}^{\infty} \quad \text{on integrating with respect to } x$$

$$= 0, \quad \text{as required.}$$

An alternative justification for the Fourier formulae for both series and integrals is obtained by considering the theory of groups. (The rest of this section requires a small knowledge of group theory and will not be made use of at any other point of the book: some readers may choose to go at once to Section 5.2. The group theoretic viewpoint is, however, fundamental to many modern generalizations of Fourier theory.)

Definition 5.2 An *Abelian group* is a set G of elements together with a binary operation (which we shall denote by $+$) having the properties:

   (i) for each x, y in G, $x + y$ is in G;
   (ii) for each x, y, z in G, $(x + y) + z = x + (y + z)$;
  (iii) for each x, y in G. $x + y = y + x$;
  (iv) there is an element 0 in G such that $x + 0 = x$ for all x in G;
   (v) for each x in G, there is an element $-x$ in G such that $x + (-x) = 0$.

Obvious examples of groups are the real numbers $\mathbb{R}$ and the integers $\mathbb{Z}$ under addition. Another example which will be very important to us is as follows: take a real number $a > 0$ and partition the real numbers into equivalence classes by declaring x to be equivalent to y if $x - y$ is an integer multiple of a. These equivalence classes themselves form a group when addition of classes is defined element by element; this group we shall denote by $\mathbb{T}_a$.

This group $\mathbb{T}_a$ may be regarded more concretely by choosing from each equivalence class its unique representative in the interval $[0, a)$ and by adding representatives 'modulo a', i.e. the sum of x and y is to be $x + y$ or $x + y - a$, whichever is again in $[0, a)$. $\mathbb{T}_a$ may also be identified with the unit circle S_1, which is the set of complex numbers of modulus one under multiplication, by the mapping $x \to e^{2\pi i x / a}$. Evidently the structure of $\mathbb{T}_a$ is independent of the value of a, and it is usual to ta $a = 1$ or $a = 2\pi$ and drop the suffix.

Both $\mathbb{T}_a$ and $\mathbb{R}$ h .....portant subsets which are groups in their own right, such as the set of all rational numbers, or the rational numbers of the form m/p^n where p is a fixed prime, and m, n are any integers. For the moment we shall consider only $\mathbb{R}$, $\mathbb{Z}$, and $\mathbb{T}$.

Definition 5.3 Let $(G, +)$ be an Abelian group, and let χ be a mapping of G to the circle group S_1 as defined above. We say that χ is a *group character of* G if $\chi(x + y) = \chi(x) \cdot \chi(y)$ for all x, y in G. (In group theoretic terms χ is a group homomorphism of G into S_1).

Lemma 5.4 (i) The set of all group characters of a given group $(G, +)$ form another group Γ with the operation of multiplication: $(\chi_1 + \chi_2)(x) = \chi_1(x) \cdot \chi_2(x)$.
  (ii) Each character of $\mathbb{Z}$ has the form $n \to e^{inx}$ for a unique x in $[0, 2\pi)$.
  (iii) Each continuous character of $\mathbb{T}$ has the form $x \to e^{inx}$ for a unique n in $\mathbb{Z}$.
  (iv) Each continuous character of $\mathbb{R}$ has the form $x \to e^{ixy}$ for a unique y in $\mathbb{R}$.
  (A continuous character is, of course, a function which is both a character as defined above, and a continuous function in the normal sense on $\mathbb{R}$, or in the sense applied to periodic functions (i.e. functions on $\mathbb{T}$) in section 2, Chapter 1.)

Proof. (i) is left as an easy exercise for the reader (exercise 1 at the end of the chapter).

For (ii) notice that $\mathbb{Z}$ is generated by the single element 1: hence if $\chi(1) = e^{ix}$ where x is a uniquely determined number in $[0, 2\pi)$, then $\chi(-1) = (\chi(1))^{-1} = e^{-ix}$, and for any integer n, $\chi(n) = (\chi(1))^n = e^{inx}$ as required.

In order to prove (iii) and (iv) we need the fact that if ϕ is a continuous function which is defined on some interval of the form $I = (-h, h)$ and which satisfies

$\phi(x + y) = \phi(x) + \phi(y)$ whenever x, y, and $x + y$ are in I, then ϕ is linear ($\phi(x) = cx$ for some fixed c, and all x in I). To prove this notice that the condition on ϕ implies that $\phi(nx) = n\phi(x)$ for integer n, when nx is in I. Hence we get successively

$$\phi(x) = n\phi\left(\frac{x}{n}\right) \quad \text{and} \quad \phi\left(\frac{kx}{n}\right) = \frac{k}{n}\phi(x)$$

for integers k and n. The continuity of ϕ now shows that if we choose a sequence of rational numbers k/n which approach a real number t, then $\phi(tx) = t\phi(x)$, provided both x and tx are in I. It follows that if we fix some x_0 in I, then

$$\phi(x) = \frac{x}{x_0}\phi(x_0)$$

for all x in I, which proves that ϕ is linear.

Now let χ be any continuous character of $\mathbb{R}$: the continuity at $x = 0$ shows that $\chi(x) \to 1$ as $x \to 0$ and so for $|x| < h$, say, $|\chi(x) - 1| < 1$ and so $\log \chi(x)(= \phi(x)$ say) may be unambiguously defined. But then $\phi(x + y) = \log \chi(x + y) = \log \chi(x) + \log \chi(y) = \phi(x) + \phi(y)$ if all of x, y, and $x + y$ are in $(-h, h)$, so the above result shows that ϕ is linear; i.e. $\log \chi(x) = \phi(x) = cx$ for some fixed c and $-h < x < h$, and hence $\chi(x) = e^{cx}$. But x is real and $|\chi(x)| = 1$, so c must be of the form iy for some real y. If x is not in the interval $(-h, h)$, choose an integer n such that $|x/n| < h$, and so $\chi(x) = (\chi(x/n))^n = (e^{cx/n})^n = e^{ixy}$ as required. Since different values of y give different characters, the uniqueness is obvious and this completes the proof of (iv). The proof of (iii) is similar, the only difference being that the continuity of χ on $\mathbb{T}$ requires that $\chi(0) = \chi(2\pi)$ so the character $\chi(x) = e^{ixy}$ found above for $\mathbb{R}$ must have an integer value of y.

The group Γ of characters of a given group G (restricted by the requirement of continuity when appropriate as in the case of $\mathbb{R}$ and $\mathbb{T}$ above) is called the *dual group of G*: thus we have identified the dual group of $\mathbb{T}$, $\mathbb{Z}$, and $\mathbb{R}$ as $\mathbb{Z}$, $\mathbb{T}$, and $\mathbb{R}$ respectively. In each case we see that the Fourier transform of a function f on G is a function $\hat{f}$ on Γ defined by

$$\hat{f}(n)(= c_n) = \frac{1}{2\pi}\int_0^{2\pi} f(x)e^{-inx}\,\mathrm{d}x,$$

for a function f defined on $\mathbb{T}$ or

$$\hat{f}(x) = \sum_{-\infty}^{\infty} f_n e^{-inx},$$

for a sequence (f_n) defined on $\mathbb{Z}$, and $0 \leqslant x \leqslant 2\pi$, or $\hat{f}(y) = \int_{-\infty}^{\infty} f(x)e^{-2\pi ixy}\,\mathrm{d}x$, for a function f on $\mathbb{R}$, where each of the right-hand sides has the form of a sum or integral of the function multiplied by a suitable character.

Thus each version of the Fourier transform whether by series or integrals is in fact a special case of a general process which can be defined for a wide class of groups. This is a direction which we cannot pursue here – the reader is referred to

Appendix C for more advanced books in this area. Hopefully we have shown, however, that the Fourier integral as defined in equations (5.4) and (5.5) is indeed the correct version to be used for non-periodic functions on $\mathbb{R}$, and we may now proceed to its systematic study.

5.2 THE FOURIER INTEGRAL

Throughout the remainder of this chapter we shall be concerned with real- or complex-valued FV-functions, as defined in Definition A.13. That is, each function f will be defined on the real line, and have the following properties:

(i) On each bounded interval $[a, b]$, f will be an FC-function, being bounded and having only a finite set of discontinuities;

(ii) $\int_a^b |f(x)|\, dx$ approaches a finite limit as $a \to -\infty$, and $b \to \infty$.

In particular, for any FV-function $f, \int_{-\infty}^{\infty} f(x)\, dx$ may be defined as $\lim_{a \to \infty, b \to -\infty} \int_a^b f(x)\, dx$. On the whole real line an FV-function may have infinitely many discontinuities and be unbounded, as example (iv) below shows.

Examples 5.5 (i) Let

$$f(x) = \begin{cases} x^{-2} & \text{for } x \geqslant 1 \\ 0 & \text{for } x < 1 \end{cases}$$

Then f is as FV-function and $\int_{-\infty}^{\infty} f(x)\, dx = 1$.

For f is continuous except at $x = 1$, $f(x) \leqslant 1$ for all x, and assuming $a < 1 < b$ we have

$$\int_a^b f(x)\, dx = \int_1^b \frac{dx}{x^2} = 1 - 1/b \to 1 \text{ as } b \to \infty.$$

More generally, if $f(x) = x^{-p}$ for $x \geqslant 1$, and 0 elsewhere, then f is an FV-function if $p > 1$ (but not if $p \leqslant 1$) and $\int_{-\infty}^{\infty} f(x)\, dx = (p-1)^{-1}$.

(ii) Let

$$f(x) = \begin{cases} x^{-\frac{1}{2}} & \text{for } 0 < x \leqslant 1 \\ 0 & \text{elsewhere} \end{cases}$$

Then f is not an FV-function, being unbounded on $[0, 1]$. (This is despite the fact that f is improperly integrable over $(0, 1)$.)

(iii) Let $f(x) = \cos cx$, $\sin cx$, or e^{icx} for some real number c. Then in each case f is bounded and continuous on the real line, but $\int_a^b f(x)\, dx$ has no limit as $a \to -\infty$, $b \to \infty$, and f is not an FV-function.

(iv) Let f be defined by

$$f(x) = \begin{cases} n & \text{if } n \leqslant x < n + n^{-3}, \quad \text{for some } n = 1, 2, 3, \ldots. \\ 0 & \text{otherwise} \end{cases}$$

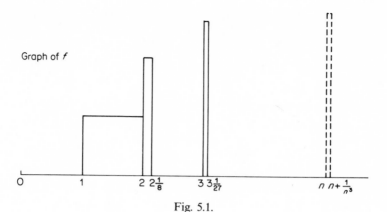

Graph of f

Fig. 5.1.

Then f has a discontinuity at each point $x = n$, or $n + n^{-3}$, and $f(n) = n$, so f is unbounded on the whole line (Fig. 5.1) However, f is positive and $\int_n^{n+1} f(x)\,dx = \int_n^{n+n^{-3}} n\,dx = n^{-2}$, so that f is an FV-function and $\int_{-\infty}^{\infty} f(x)\,dx = \sum_{n=1}^{\infty} n^{-2} = \pi^2/6$.

(v) Let

$$f(x) = \begin{cases} \dfrac{\sin x}{x} & \text{for } x \neq 0 \\ 1 & \text{if } x = 0. \end{cases}$$

It is shown in Appendix B that $\lim_{a \to \infty} \int_0^a f(x)\,dx$ exists (and is equal to $\pi/2$). However, f is not an FV-function since $\int_0^a |f(x)|\,dx$ does not have a finite limit as $a \to \infty$: for instance, on the interval $(k\pi, (k+1)\pi)$, we have

$$\frac{1}{x} \geqslant \frac{1}{(k+1)\pi}$$

so that

$$\int_{k\pi}^{(k+1)\pi} \left| \frac{\sin x}{x} \right| dx \geqslant \frac{1}{(k+1)\pi} \int_{k\pi}^{(k+1)\pi} |\sin x|\,dx = \frac{2}{(k+1)\pi},$$

and

$$\sum_{1}^{\infty} \frac{1}{k+1}$$

is divergent.

We see at once that if f is an FV-function then so is g where $g(x) = f(x)e^{-2\pi ixy}$ for real y, since the exponential factor is bounded and continuous. Hence the following definition of the Fourier transform of an FV-function is meaningful.

Definition 5.6 Let f be an FV-function on the real line. Then the *Fourier transform* $\hat{f}$ of f is defined by

$$\hat{f}(y) = \int_{-\infty}^{\infty} f(x)e^{-2\pi ixy}\,dx, \text{ for real values of } y.$$

It should be observed here that the notation is not quite standard, since the integrals

$$\int_{-\infty}^{\infty} f(x)e^{-ixy}\,dx \quad \text{and} \quad \frac{1}{\sqrt{2\pi}}\int_{-\infty}^{\infty} f(x)e^{-ixy}\,dx$$

are also commonly used to define the Fourier transform. Our choice is made on the grounds of symmetry (see equations (5.4) and (5.5)) and its close links with Example 5.1.

The Fourier transform has a number of important formal properties, some of which are immediate from the definition and are listed as Lemma 5.7, while those which require deeper consideration are given in Theorem 5.8.

Lemma 5.7 Let f, g be FV-functions. Then
 (i) $\widehat{f + g}(y) = \hat{f}(y) + \hat{g}(y)$;
 (ii) $\widehat{\alpha f}(y) = \alpha\hat{f}(y)$ for any complex number α;
 (iii) if $g(x) = \overline{f(x)}$, so g is the complex conjugate of f, then $\hat{g}(y) = \overline{\hat{f}(-y)}$;
 (iv) if $f_h(x) = f(x - h)$ so f_h is the translate of f by h, as in Definition A.15, then $\hat{f}_h(y) = e^{-2\pi ihy}\hat{f}(y)$;
 (v) $|\hat{f}(y)| \leqslant \int_{-\infty}^{\infty}|f(x)|\,dx$ for all y;
 (vi) if $h(x) = f(\lambda x)$ for a real number $\lambda \neq 0$, then $\hat{h}(y) = 1/\lambda f(y/\lambda)$.

Proof All of these are immediate deductions from Definition 5.6.
 For example, in (iii) where $g(x) = \overline{f(x)}$, then

$$\hat{g}(y) = \int_{-\infty}^{\infty} \overline{f(x)}e^{-\pi ixy}\,dx = \left(\int_{-\infty}^{\infty} f(x)e^{2\pi ixy}\,dx\right)^{-}$$

$$= (\hat{f}(-y))^{-} \text{ as required,}$$

while in part (vi),

$$\hat{h}(y) = \int_{-\infty}^{\infty} f(\lambda x)e^{-2\pi ixy}\,dx$$

$$= \int_{-\infty}^{\infty} f(t)e^{-2\pi ity/\lambda}\,dt/\lambda \quad \text{on putting } \lambda x = y$$

$$= \frac{1}{\lambda}\hat{f}(y/\lambda), \quad \text{as stated}$$

Theorem 5.8 Let f be an FV-function. Then

(i) if also $F(x) = \int_{-\infty}^{x} f(t) \, dt$ is an FV-function, then

$$\hat{F}(y) = \frac{1}{2\pi i y} \hat{f}(y);$$

(ii) if $g(x) = xf(x)$ is an FV-function then $\hat{f}$ is differentiable and

$$\frac{d}{dy} \hat{f}(y) = -2\pi i \hat{g}(y);$$

(iii) $\hat{f}$ is uniformly continuous, and $\hat{f}(y) \to 0$ as $|y| \to \infty$.
(The latter statement is of course the Riemann–Lebesgue lemma on the real line.)

Proof (i) Notice that since f is an FV-function, $\int_{-\infty}^{x} f(t) \, dt$ must have a limit as $x \to +\infty$ and since also F is an FV-function this limit must be zero, so that $\hat{f}(0) = \int_{-\infty}^{\infty} f(x) \, dx = 0$. Then

$$\hat{F}(y) = \int_{-\infty}^{\infty} F(x) e^{-2\pi i x y} \, dx = -\frac{1}{2\pi i y} [e^{-2\pi i x y} F(x)]_{-\infty}^{\infty} + \frac{1}{2\pi i y} \int_{-\infty}^{\infty} e^{-2\pi i x y} f(x) \, dx$$

on integrating by parts. The integrated term is zero since we have just shown that $F(x) \to 0$ as $x \to \infty$ (and as $x \to -\infty$ by the definition of F), and the integral is

$$\frac{1}{2\pi i y} \hat{f}(y).$$

as required
(ii) For any real y, $y + h$, we have

$$\frac{1}{h}(\hat{f}(y + h) - \hat{f}(y)) = \int_{-\infty}^{\infty} f(x) \frac{1}{h}(e^{-2\pi i x h} - 1) e^{-2\pi i x y} \, dx$$

To estimate the bracketed term we notice that for real θ, we have $|e^{i\theta} - 1| = |\int_{0}^{\theta} e^{i\phi} \, d\phi| \leqslant |\int_{0}^{\theta} d\phi| = |\theta|$, and hence that

$$\left| \frac{1}{h}(e^{-2\pi i x h} - 1) \right| \leqslant |2\pi x|.$$

Since $xf(x)$ is an FV-function, we know that given $\varepsilon > 0$, we can find a value of a such that $\int_{|x| \leqslant a} |2\pi x f(x)| \, dx < \varepsilon$. However, on the bounded set $[-a, a]$, the function $1/h(e^{-2\pi i x h} - 1)$ tends *uniformly* to $-2\pi i x$ as $h \to 0$. Combining these two facts, we see that

$$\frac{1}{h}\{(\hat{f}(y + h) - \hat{f}(y))\} \to -2\pi i \int_{-\infty}^{\infty} x f(x) e^{-2\pi i x y} \, dx \quad \text{as} \quad h \to 0,$$

which is the required result.
(iii) To show that $\hat{f}$ is uniformly continuous we put

$$\hat{f}(y + h) - \hat{f}(y) = \int_{-\infty}^{\infty} f(x)(e^{-2\pi i h x} - 1) e^{-2\pi i x y} \, dx,$$

and deduce that $|\tilde{f}(y+h) - \tilde{f}(y)| \leq \int_{-\infty}^{\infty} |f(x)||e^{-2\pi ihx} - 1|\,dx$: the remainder of the argument is like (ii) above, since $e^{-2\pi ihx} - 1 \to 0$ as $h \to 0$, and the continuity is uniform since this estimate is independent of y.

To show that $\tilde{f}(y) \to 0$ as $|y| \to \infty$, we again choose a real number a with $\int_{|x| \geq a} |f(x)|\,dx \leq \varepsilon$, and put

$$f_1 = \begin{cases} f & \text{on } [-a, a], \\ 0 & \text{elsewhere.} \end{cases}$$

Then for $y \neq 0$,

$$\hat{f}_1(y) = \int_{-a}^{a} f(x)e^{-2\pi ixy}\,dx$$

$$= \int_{-a+h}^{a+h} f(u-h)e^{-2\pi i(u-h)y}\,du, \qquad \text{putting } x = u - h$$

$$= -\int_{-a+h}^{a+h} f_h(u)e^{-2\pi iuy}\,du, \qquad \text{if we choose } h = 1/2|y|,$$

$$= -\int_{-a}^{a} f_h(x)e^{-2\pi ixy}\,dx - \int_{a}^{a+h} + \int_{-a}^{a+h}$$

(where the integrand is the same each time).

Taking the average of the first and last expressions for $\hat{f}_1(y)$ we see that

$$|\hat{f}_1(y)| \leq \frac{1}{2}\int_{-a}^{a} |f(x) - f_h(x)|\,dx + \frac{1}{2}\int_{a}^{a+h} |f_h(x)|\,dx + \frac{1}{2}\int_{-a}^{-a+h} |f_h(x)|\,dx.$$

The second and third terms are bounded by Mh, where $M = \sup|f(x)|$, and the first tends to zero as $h \to 0$ (and hence as $|y| \to \infty$) by Corollary A.18; and hence $\hat{f}_1(y) \to 0$ as $|y| \to \infty$ (observe that we have effectively reproved exercise 7, Ch. 1). But

$$|\tilde{f}(y)| = \left| \int_{|x| \geq a} f(x)e^{-2\pi ixy}\,dx \right| + \left| \int_{-a}^{a} f(x)e^{-2\pi ixy}\,dx \right|$$

$$\leq \varepsilon + |\hat{f}_1(y)| < 2\varepsilon \qquad \text{for large enough } y.$$

and the result is proved.

Notice that part (iii) of this result gives yet another reason why we should not expect the inversion formula

$$f(x) = \int_{-\infty}^{\infty} \hat{f}(y)e^{2\pi ixy}\,dy,$$

which represents $f(-x)$ as the Fourier transform of $\hat{f}$, to hold generally, since a Fourier integral must be uniformly continuous and tend to zero at $\pm\infty$, neither of which properties hold for a general FV-function.

Before taking the theory any further, we look at a few examples where Fourier transforms can be explicitly and easily calculated.

Examples 5.9 (i) Let $f(x) = e^{-ax^2}$, where a is a positive number. Then

$$\hat{f}(y) = \int_{-\infty}^{\infty} e^{-ax^2 - 2\pi ixy}\, dx$$

$$= \sqrt{\frac{\pi}{a}} \int_{-\infty}^{\infty} e^{-\pi t^2 - 2\pi ity\sqrt{\pi/a}}\, dy, \qquad \text{putting } ax^2 = \pi t^2$$

$$= \sqrt{\frac{\pi}{a}}\, e^{-\pi(y^2\pi/a)}, \qquad \text{by Example 5.1}$$

$$= \sqrt{\frac{\pi}{a}}\, e^{-\pi^2 y^2/a}.$$

(Example 5.1 is of course the special case when $a = \pi$. The differentiation argument used there is justified by Theorem 5.8(ii) above.) The reader is invited to extend the above result to the case when a is complex, and Re $a > 0$.

(ii) Let

$$f(x) = \begin{cases} 1 & \text{on } [a, b], \\ 0 & \text{elsewhere.} \end{cases}$$

then

$$\hat{f}(y) = \int_a^b e^{-2\pi ixy}\, dx = -\frac{1}{2\pi iy}[e^{-2\pi ixy}]_a^b = \frac{e^{-2\pi iay} - e^{-2\pi iby}}{2\pi iy}.$$

In particular, if the interval has the symmetric form $[-a, a]$, then

$$\hat{f}(y) = \frac{\sin(2\pi ay)}{\pi y}.$$

In either case, although $\hat{f}$ is continuous and tends to zero as $|y| \to \infty$ (as it must by Theorem 5.8(iii)) $\hat{f}$ is not an FV-function (as is pointed out in Examples 5.5(v)). The possibility which this example illustrates, namely that $\hat{f}$ may not be integrable, has to be borne in mind all the time when dealing with Fourier integrals, and is responsible for many of the problems, and subtleties which arise.

(iii) Let $f(x) = 1/(t^2 + x^2)$, for a non-zero real number t. Then

$$\hat{f}(y) = \int_{-\infty}^{\infty} \frac{e^{-2\pi ixy}}{t^2 + x^2}\, dx = 2\int_0^{\infty} \frac{\cos 2\pi xy}{t^2 + x^2}\, dx = \frac{2}{t}\int_0^{\infty} \frac{\cos 2\pi tyu}{1 + u^2}\, du,$$

putting $x = tu$, if $t > 0$, and hence it is sufficient to evaluate the integral

$$g(m) = \int_0^{\infty} \frac{\cos mu}{1 + u^2}\, du, \text{ for real values of } m.$$

The value of $g(m)$ is known to be $(\pi/2)e^{-|m|}$ and is usually found using the theory of residues: we outline briefly a real-variable method for finding its value. Suppose that $m > 0$, and integrate the integral which defines $g(m)$ by parts to obtain

$$g(m) = \frac{1}{m}\int_0^\infty \sin mu \frac{2u}{(1+u^2)^2}\,du.$$

This may be differentiated with respect to m to give

$$(mg(m))' = mg'(m) + g(m) = \int_0^\infty \frac{2u^2}{(1+u^2)^2}\cos mu\,du$$

$$= 2\int_0^\infty \left(1 - \frac{1}{1+u^2}\right)\frac{\cos mu}{1+u^2}\,du$$

$$= 2g(m) - 2\int_0^\infty \frac{\cos mu}{(1+u^2)^2}\,du,$$

$$\text{or}\quad mg'(m) - g(m) = -2\int_0^\infty \frac{\cos mu}{(1+u^2)^2}\,du$$

If we differentiate again, we get

$$mg''(m) + g'(m) - g'(m) = +2\int_0^\infty \frac{u\sin mu}{(1+u^2)^2}\,du$$

$$= mg(m), \text{ as noted above.}$$

Hence $g''(m) - g(m) = 0$, so that $g(m)$ must have the form $Ae^m + Be^{-m}$ for suitable constants A and B. But $g(m)$ is a Fourier integral and so tends to zero as $m \to 0$, so A must be zero: also

$$g(m) \to \int_0^\infty \frac{du}{1+u^2} = \frac{\pi}{2}$$

when $m \to 0$, so $B = \pi/2$.

Hence $g(m) = \pi e^{-m}/2$ for $m > 0$, and since plainly $g(m) = g(-m)$,

$$g(m) = \frac{\pi}{2}e^{-|m|} \qquad \text{for all real } m.$$

It follows that for

$$f(x) = \frac{1}{t^2 + x^2},$$

$$\hat{f}(y) = \frac{2}{t}g(2\pi ty) \qquad \text{for } t > 0$$

$$= \frac{\pi}{t}e^{-2\pi t|y|},$$

and hence for any real $t \neq 0$,

$$\hat{f}(y) = \frac{\pi}{|t|}e^{-2\pi|ty|}.$$

(iv) Let $f(x) = e^{-a|x|}$ for real x, and $a > 0$.
Then

$$\hat{f}(y) = \int_{-\infty}^{\infty} e^{-a|x|} e^{-2\pi i x y} \, dx$$

$$= \int_{-\infty}^{0} e^{(a - 2\pi i y)} \, dx + \int_{0}^{\infty} e^{-(a + 2\pi i y)x} \, dx$$

$$= (a - 2\pi i y)^{-1} + (a + 2\pi i y)^{-1}$$

$$= \frac{2a}{a^2 + 4\pi^2 y^2} \cdot$$

The observant reader will have noticed already that parts (iii) and (iv) of the preceding example give another instance of the inversion formulae (5.4) and (5.5) in action.
For if

$$f(x) = \frac{1}{t^2 + x^2}, \quad \text{then } \hat{f}(y) = \frac{\pi}{|t|} e^{-2\pi|ty|} \quad \text{by (iii)}.$$

Then the inversion integral

$$\int_{-\infty}^{\infty} \hat{f}(y) e^{2\pi i x y} \, dy = \frac{\pi}{|t|} \int_{-\infty}^{\infty} e^{2\pi|ty|} e^{2\pi i x y} \, dx$$

$$= \frac{\pi}{|t|} \frac{2(2\pi|t|)}{(2\pi|t|)^2 + 4\pi^2 x^2} \quad \text{by (iv)}$$

$$= \frac{4\pi^2}{4\pi^2(t^2 + x^2)} = \frac{1}{t^2 + x^2},$$

as required to verify the formula.

Our next objective is to define the notion of convolution for FV-functions: unfortunately there is a technical difficulty in that the natural definition $h(x) = (f*g)(x) = \int_{-\infty}^{\infty} f(t)g(x - t) \, dt$ is not defined for all FV-functions and all real values of x. (To see this, notice that Example 5.5(iv) gives an example of an FV-function f for which f^2 is not an FV-function. Consequently if $g(x) = f(-x)$, then the above integral for $f*g(0) = \int_{-\infty}^{\infty} f(t)g(-t) \, dt$ does not exist.) In order to avoid this problem we shall restrict the convolution operation to the case when at least one of the functions involved is bounded; that is, there exists a number M such that $|f(x)| \leqslant M$ for all real x.

(The more advanced theory using the Lebesgue integral requires only that f and g are defined almost everywhere, and in this case when f and g are integrable, then $h = f*g$ is defined almost everywhere and is also integrable.)

Lemma 5.10 Let f, g be FV-functions one of which is bounded. Then the integral $h(x) = \int_{-\infty}^{\infty} f(t)g(x - t) \, dt$ exists for all real x and defines h as a continuous FV-function.

Proof Since both f and g have only a finite number of discontinuities and are bounded on any interval of finite length, the same is true of $\phi(t) = f(t)g(x - t)$.

It follows that $\int_a^b \phi(t)\,dt$ exists for any real values of a and b; moreover, if say, g is bounded, $|g(x)| \leqslant M$, then $\int_a^b |\phi(t)|\,dt \leqslant M\int_a^b |f(t)|\,dt$ and so ϕ is itself an FV-function. (Since clearly $\int_{-\infty}^{\infty} f(t)g(x - t)\,dt = \int_{-\infty}^{\infty} f(x - u)g(u)\,du$ putting $t = x - u$, the same is true if it is f which is bounded.) It follows that h is well defined by either of the integrals $h(x) = \int_{-\infty}^{\infty} f(t)g(x - t)\,dt = \int_{-\infty}^{\infty} f(x - u)g(u)\,du$.

The continuity (and indeed the uniform continuity) of h follows by putting

$$|h(x) - h(y)| = \left| \int_{-\infty}^{\infty} f(t)[g(x - t) - g(y - t)]\,dt \right|$$

$$\leqslant M \int_{-\infty}^{\infty} |g(x - t) - g(y - t)|\,dt$$

(assuming f is the bounded factor) and applying Corollary A.18.

The fact that h is an FV-function now follows on noting that $\int_a^b |h(x)|\,dx \leqslant \int_{-\infty}^{\infty} |f(t)| \int_a^b |g(x - t)|\,dx\,dt$ is bounded above by $\int_{-\infty}^{\infty} |f(t)|\,dt$, $\int_{-\infty}^{\infty} |g(u)|\,du$.

Lemma 5.10 allows us to make the claims embodied in the following definition

Definition 5.11 Let f, g be FV-functions of which one is bounded. Then the *convolution* $(f*g)(x) = \int_{-\infty}^{\infty} f(t)g(x - t)\,dt$ is a continuous FV-function, and the operation of convolution is commutative and distributive over addition. For the associative property to apply $(f*(g*h) = (f*g)*h)$ it is sufficient that (at least) the middle factor g should be bounded.

The convolution product has formal properties analogous to those in Chapter 1, making it a particularly useful tool for our purposes. In particular, we have the following result.

Lemma 5.12 Let f, g be FV-functions of which one is bounded, as in Definition 5.11, and let $h = f*g$. Then

$$\hat{h}(y) = \hat{f}(y)\hat{g}(y) \qquad \text{for all real } y.$$

Proof By definitions 5.6 and 5.11 we have

$$\hat{h}(y) = \int_{-\infty}^{\infty} h(x)e^{-2\pi i x y}\,dx$$

$$= \int_{-\infty}^{\infty} \left(\int_{-\infty}^{\infty} f(t)g(x - t)\,dt \right) e^{-2\pi i x y}\,dx$$

$$= \int_{-\infty}^{\infty} f(t) \left(\int_{-\infty}^{\infty} g(x - t)e^{-2\pi i(x - t)y}\,dx \right) e^{-2\pi i t y}\,dt,$$

on interchanging the order of the integration (see the note at the end of this section)

$$= \int_{-\infty}^{\infty} f(t)(\hat{g}(y))e^{-2\pi i t y}\,dt$$

$$= \hat{f}(y)\hat{g}(y) \qquad \text{as stated.}$$

The following result will be fundamental for our study of the inversion integral in the next section.

Theorem 5.13 Let f, ϕ be FV-functions, and suppose that $\hat{\phi}$ is also an FV-function. Then $\phi(-y)\hat{f}(y)$ is an FV-function, $f*\hat{\phi}$ is defined, and we have that for all real x,

$$(f*\hat{\phi})(x) = \int_{-\infty}^{\infty} \phi(-y)\hat{f}(y)e^{2\pi i x y}\,dy.$$

Proof $\hat{\phi}$ is a continuous and bounded FV-function, so that $f*\hat{\phi}$ is defined as in Definition 5.11. Similarly, $\hat{f}$ is bounded and continuous so that $\phi(-y)\hat{f}(y)$ is an FV-function. We then have

$$(f*\hat{\phi})(x) = \int_{-\infty}^{\infty} f(t)\hat{\phi}(x-t)\,dt$$

$$= \int_{-\infty}^{\infty} f(t)\int_{-\infty}^{\infty} \phi(u)e^{-2\pi i(x-t)u}\,du\,dt$$

$$= \int_{-\infty}^{\infty} \phi(u)e^{-2\pi i x u}\left(\int_{-\infty}^{\infty} f(t)e^{2\pi i t u}\,dt\right)du$$

$$= \int_{-\infty}^{\infty} \phi(u)e^{-2\pi i x u}\hat{f}(-u)\,du, \text{ as required.}$$

Note Throughout the preceding discussion a crucial role was played by Fubini's theorem, which enables us to justify interchanging the order of integration when needed. The theorem is proved in Theorem A.33 *et seq.*

5.3 SUMMABILITY AND INVERSION OF THE FOURIER INTEGRAL

In the preceding section we examined the properties of the Fourier integral, seeing in Theorem 5.8, for example, how properties of f were reflected in those of $\hat{f}$. We now have to tackle the major theoretical problem, which is to reconstruct f from a knowledge of its Fourier transform. Section 5.1 suggested that we might have simply $f(x) = \int_{-\infty}^{\infty} \hat{f}(y)e^{2\pi i x y}\,dy$, and the last part of Theorem 5.17 below gives some conditions under which this is true. However, it cannot be valid for all FV-functions f, since we already have an example (5.9(ii)) when $\hat{f}$ need not be integrable. Moreover, when $\hat{f}$ is integrable, the integral $\int_{-\infty}^{\infty} \hat{f}(y)e^{2\pi i x y}\,dy$ is a continuous function of x, so that continuity of f is another condition which must be satisfied for the inversion formula to be valid in this form.

We must therefore be more careful in our use of the inversion formula; in particular, we shall interpret the above integral in certain cases by a summability process, as was done in Chapter 3 for Fourier series. There are at least three natural summability processes which can be used to assist with the problem of Fourier inversion: we shall use the one which is most directly analogous to the Poisson summability used in Chapter 3, and leave the others to be investigated as exercises (Definition 5.14 and exercises 5 and 6 at the end of this chapter).

Our method of attack on this problem will be as follows. We consider the result established at the end of the last section:

$$(f * \hat{\phi})(x) = \int_{-\infty}^{\infty} \phi(-y)\hat{f}(y)e^{2\pi i xy}\,dy,$$

and we substitute for ϕ a particular family (ϕ_t) of functions (where t is a real parameter) which has the property that for each real y, $\phi_t(y) \to 1$ as $t \to 0$. The right-hand side of this equation thus becomes (in a sense to be investigated) the integral $\int_{-\infty}^{\infty} \hat{f}(y)e^{2\pi i xy}\,dy$, and our aim will be to show that at the same time, $f * \hat{\phi}_t$ approaches f as $t \to 0$. We begin by defining the particular family (ϕ_t) which we shall use.

Definition 5.14 (i) Let $\phi_t(x) = e^{-t|x|}$ for real $t > 0$ and real x. We shall say that the integral $\int_{-\infty}^{\infty} f(x)dx$ is *P-summable*, and that its (P-) value is A if the limit

$$\lim_{t \to 0} \int_{-\infty}^{\infty} e^{-t|x|} f(x)\,dx \text{ exists and is equal to } A.$$

We write (P) $\int_{-\infty}^{\infty} f(x)\,dx = A$ when this occurs.

We shall show below that this limit may exist when the ordinary integral $\int_{-\infty}^{\infty} f(x)\,dx$ does not (for instance when $f(x) = \cos x$), but that whenever $\int_{-\infty}^{\infty} f(x)\,dx$ does exist, it is P-summable, to the same value. Hence we can say that P-summability is both compatible with ordinary convergence, and strictly includes it.

(ii) Let $\phi_t(x) = e^{-tx^2}$ for $t > 0$ and real x. We may say, in the same way as in (i), that $\int_{-\infty}^{\infty} f(x)\,dx$ is *G-summable*, with value A if $\lim_{t \to 0} \int_{-\infty}^{\infty} e^{-tx^2} f(x)\,dx$ exists and is equal to A.

Similarly, if

$$\phi_t(x) = \begin{cases} 1 - t|x| & \text{for } 0 \leqslant |x| \leqslant t^{-1} \\ 0 & \text{if } |x| > t^{-1} \end{cases}$$

we say that $\int_{-\infty}^{\infty} f(x)dx$ is *C-summable* with value A if $\lim_{t \to 0} \int_{-1/t}^{1/t} (1 - t|x|)f(x)dx$ exists and is equal to A.

The letters P, G, and C are for Poisson, Gauss, and Cesàro respectively.

Examples 5.15 Let $f(x) = \sin mx$ for real $m \neq 0$: we shall find the P-value of

$\int_0^\infty \sin mx \, dx$. Since $\int_0^a |\sin mx| \, dx \to \infty$ with a, this function is not integrable in the usual sense.

For $t > 0$ we see that

$$\int_0^\infty e^{-tx} \sin mx \, dx = \frac{1}{2i} \int_0^\infty \{e^{-(t-im)x} - e^{-(t+im)x}\} \, dx$$

$$= \frac{1}{2i} \left\{ \frac{1}{t-im} + \frac{1}{t+im} \right\}$$

$$= \frac{m}{t^2 + m^2}$$

which tends to $1/m$ as $t \to 0$.

Hence we have shown that $(P)\int_0^\infty \sin mx \, dx = 1/m$ for real $m \neq 0$.

Exercise 8 at the end of the chapter gives some related integrals which can be found in this way.

We now prove the compatibility of P-summability with ordinary convergence.

Theorem 5.16 Let f be an F-function: more generally, let f be any function, which is FC on every bounded interval and for which both limits

$$\lim_{a \to \infty} \int_0^a f(x) \, dx = l_1 \qquad \text{and} \qquad \lim_{b \to -\infty} \int_b^0 f(x) \, dx = l_2$$

exist. Then

$$\lim_{t \to 0} \int_{-\infty}^\infty e^{-t|x|} f(x) \, dx \text{ exists and is equal to } l_1 + l_2 \left(= \int_{-\infty}^\infty f(x) \, dx \right).$$

Proof It is obviously sufficient to consider $\int_0^\infty f(x) \, dx$ alone. We have then to show that if $F(x) = \int_0^x f(t) \, dt \to l_1$ as $x \to \infty$, then also $\int_0^\infty e^{-tx} f(x) \, dx \to l_1$ as $t \to 0$.

The existence of $\int_0^\infty e^{-tx} f(x) \, dx$ (which is not quite immediate from the existence of $l_1 = \lim_{x \to \infty} \int_0^x f(t) \, dt$) follows on integration by parts, since for any $a > 0$, we have

$$\int_0^a e^{-tx} f(x) \, dx = [e^{-tx} F(x)]_0^a + t \int_0^a e^{-tx} F(x) \, dx.$$

Since $t > 0$, we may let $a \to \infty$ to obtain

$$\int_0^\infty e^{-tx} f(x) \, dx = t \int_0^\infty e^{-tx} F(x) \, dx,$$

where the integral on the right exists, since F is bounded.

Since $t \int_0^\infty e^{-tx} \, dx = 1$, we may subtract l_1 from both sides to obtain

$$\int_0^\infty e^{-tx} f(x)\,dx - l_1 = t\int_0^\infty e^{-tx}(F(x) - l_1)\,dx:$$

recall that $F(x) \to l_1$ as $x \to \infty$, and that we require that the left-hand side of this equation should tend to zero – hence it is sufficient to prove that the right-hand side tends to zero as $t \to 0$.

Given $\varepsilon > 0$, choose x_0 such that $|F(x) - l_1| < \varepsilon$ if $x \geqslant x_0$, and thus $|t\int_{x_0}^\infty e^{-tx}(Fx) - l_1|\,dx \leqslant \varepsilon t\int_{x_0}^\infty e^{-tx}\,dx \leqslant \varepsilon$. Suppose that M is chosen so that $|F(x) - l_1| \leqslant M$ *for all* x; we then obtain

$$\left| t\int_0^{x_0} e^{-tx}(F(x) - l_1)\right| dx \leqslant Mt\int_0^{x_0} e^{-tx}\,dx \leqslant Mtx_0 \quad \text{which is} < \varepsilon \text{ if } 0 < t < \varepsilon/Mx_0.$$

Combining these results we see that

$$\left| t\int_0^\infty e^{-tx}(F(x) - l_1)\,dx\right| \leqslant 2\varepsilon \qquad \text{if } 0 < t \leqslant \varepsilon/Mx_0,$$

and the result follows.

With these preliminary results out of the way we can proceed to our main results on the summability of Fourier transforms.

Theorem 5.17 For any bounded FV-function f, the integral $\int_{-\infty}^\infty \hat{f}(y)e^{2\pi ixy}\,dy$ is P-summable with value $f(x)$ at every point of continuity of f. More explicitly,

$$\int_{-\infty}^\infty e^{-t|y|}\hat{f}(y)e^{2\pi ixy}\,dy \to f(x) \qquad \text{as } t \to 0,$$

at every point of continuity of f.

In particular, if f is bounded and continuous, and both f and $\hat{f}$ are FV-functions, then

$$\int_{-\infty}^\infty \hat{f}(y)e^{2\pi ixy}\,dy = f(x) \qquad \text{for all } x.$$

Proof Let us write $\phi_t(y) = e^{-t|y|}$ for $t > 0$: we know from Example 5.9(iv) that

$$\hat{\phi}_t(u) = \frac{2}{t^2 + 4\pi^2 u^2}$$

and from Lemma 5.13 that

$$\int_{-\infty}^\infty e^{-t|y|}\hat{f}(y)e^{2\pi ixy}\,dy = (f * \hat{\phi}_t)(x).$$

Hence in order to prove the theorem, we have to show that $(f * \hat{\phi}_t)(x) \to f(x)$ as $t \to 0$. This argument, like that in Theorem 5.16 above, depends only on rather general properties of $\hat{\phi}_t$(compare exercise 9) as follows. Suppose that we have a family of functions $(g_t)_{t>0}$ which satisfy:

(a) $g_t(x) \geqslant 0$ for all x and t;

(b) $\int_{-\infty}^{\infty} g_t(x) \, dx = 1$ for all t;

(c) for any $\delta > 0$, $\int_{|x| \geqslant \delta} g_t(x) \, dx \to 0$ as $t \to 0$.

All of these properties are immediate for our choice of $g_t = \hat{\phi}_t$ (and also for the corresponding methods of Gauss and Cesàro). We shall show, subject (a), (b), and (c), that $f * g_t(x) \to f(x)$ at each point of continuity.

To see this, write

$$f * g_t(x) - f(x) = \int_{-\infty}^{\infty} f(x - u) g_t(u) \, du - f(x)$$

$$= \int_{-\infty}^{\infty} (f(x - u) - f(x)) g_t(u) \, du \qquad \text{by (b).}$$

Suppose that x is a point of continuity, that $\varepsilon > 0$ is given, and that $\delta > 0$ is such that $|f(x - u) - f(x)| < \varepsilon$ when $|u| < \delta$.

It follows that

$$\left| \int_{-\delta}^{\delta} (f(x - u) - f(x)) g_t(u) \, du \right| \leqslant \varepsilon \int_{-\delta}^{\delta} g_t(u) \, du < \varepsilon \qquad \text{by (a) and (b).}$$

However, we may estimate the rest of the integral for $f * g_t - f$, namely $\int_{|u| \geqslant \delta} (f(x - u) - f(x)) g_t(u) \, du$, by replacing $f(x - u) - f(x)$ by $2M$ (where $M = \sup|f|$: f is assumed bounded), and this gives us the estimate $2M \int_{|u| \geqslant \delta} g_t(u) \, du$ which can be made less than ε for t small, using (c). Combining these results we see that $f * g_t(x) \to f(x)$ as $t \to 0$ as required.

The final statement of the theorem follows from Theorem 5.16 since if $\hat{f}$ is an FV-function, then the integral

$$\int_{-\infty}^{\infty} \hat{f}(y) e^{2\pi i x y} \, dy \text{ exists and is the limit of}$$

$$\int_{-\infty}^{\infty} e^{-t|y|} \hat{f}(y) e^{2\pi i x y} \, dy \text{ as } t \to 0.$$

The following corollary embodies a part of this result, and will be needed in the next section.

Corollary 5.18   Let f be a bounded FV-function which is continuous at 0, and for which $\hat{f}(y) \geqslant 0$ for all real y.

Then $\hat{f}$ is also an FV-function, and $f(0) = \int_{-\infty}^{\infty} \hat{f}(y) \, dy$.

Proof   The hypotheses imply by Theorem 5.17 that

$$f(0) = \lim_{t \to 0} \int_{-\infty}^{\infty} e^{-t|y|} \hat{f}(y) \, dy.$$

But for a *positive* function $\hat{f}$, the existence of the limit is equivalent to $\hat{f}$ being an FV-function, and the value of the limit is then $\int_{-\infty}^{\infty} \hat{f}(y) \, dy$.

We have already have a number of examples of the inversion theorem, notably in Section 5.2, and there are others in the exercises at the end of the chapter. We conclude this section with a slightly different type of example, in which the function is continuous but vanishes outside a finite interval.

Example 5.19 Let

$$f(x) = \begin{cases} \cos^2(\pi/2x) & \text{on } [-1, 1] \\ 0 & \text{for } |x| > 1 \end{cases} \quad \text{(Fig. 5.2)}.$$

We have

$$\hat{f}(y) = \int_{-1}^{1} \cos^2\left(\frac{\pi}{2}x\right) e^{-2\pi ixy} \, dx.$$

$$= \int_{-1}^{1} \cos^2\left(\frac{\pi}{2}x\right) \cos(2\pi xy) \, dx$$

$$= 2\int_{0}^{1} \cos^2\left(\frac{\pi}{2}x\right) \cos(2\pi xy) \, dx \qquad \text{(since the integral is even)}$$

$$= \int_{0}^{1} (1 + \cos \pi x) \cos(2\pi xy) \, dx$$

$$= \int_{0}^{1} [\cos(2\pi xy) + \tfrac{1}{2}\{\cos((1+2y)\pi x) + \cos((1-2y)\pi x)\}] \, dx$$

$$= \frac{\sin 2\pi y}{2\pi y} + \frac{1}{2}\left\{\frac{\sin(1+2y)\pi}{(1+2y)\pi} + \frac{\sin(1-2y)\pi}{(1-2y)\pi}\right\},$$

the last line being valid except when $y = 0$, $\pm\frac{1}{2}$.

The expression for $\hat{f}(y)$ simplifies to give

$$\sin(2\pi y)\left\{\frac{1}{2\pi y} + \frac{1}{2\pi}\left(\frac{-1}{1+2y} + \frac{1}{1-2y}\right)\right\}$$

$$= \frac{\sin 2\pi y}{2\pi}\left\{\frac{1}{y} + \frac{4y}{1-4y^2}\right\} = \frac{\sin 2\pi y}{2\pi y(1-4y^2)}.$$

Since $\hat{f}$ is known to be continuous, the values at $y = 0$, $\pm\frac{1}{2}$ can be found by taking limits in this expression (so that $\hat{f}(0) = 1$, for instance). The graph of $\hat{f}$ is

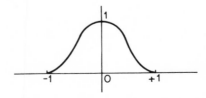

Fig. 5.2.

found from that of $\sin 2\pi y$, with the zeros at 0, $\pm\frac{1}{2}$ deleted, and the amplitude damped by an amount proportional to y^{-3}. The reader is invited to verify that f and $\hat{f}$ satisfy the hypotheses of the second part of Theorem 5.17.

5.4 SQUARE-INTEGRABLE FUNCTIONS

We now consider the analogues of the results of Section 1.4 and 1.5 and in particular Bessel's and Parseval's relations (Theorem 1.19). We again encounter the difficulty which we met in Section 5.2 in connection with the definition of convolution, namely that there are FV-functions f for which f^2 is not an FV-function.

We shall suppose for this section that f is a bounded FV-function, in which case both f and f^2 are FV-functions, and we shall show that $(\hat{f})^2$ is also an FV-function (though $\hat{f}$ may not be). We shall also derive the striking analogue of Bessel's equation $\int_{-\infty}^{\infty}|f(x)|^2\,dx = \int_{-\infty}^{\infty}|\hat{f}(y)|^2\,dy$, which shows that the Fourier transform is an *isometric* mapping if we use these integrals as our measure of the size of f.

The crucial concepts for this section are those of convolution (Definition 5.11), Corollary 5.18 and the fact (Lemma 5.7(iii)) that if $g(x) = \overline{f(-x)}$ then $\hat{g}(y) = \overline{\hat{f}(y)}$. We begin by establishing the above analogue of Bessel's equation.

Theorem 5.20 Let f be a bounded FV-function. Then $(\hat{f})^2$ is an FV-function and $\int_{-\infty}^{\infty}|f(x)|^2\,dx = \int_{-\infty}^{\infty}|\hat{f}(y)|^2\,dy$.

Proof We put $g(x) = \overline{f(-x)}$ and note as above that $\hat{g}(y) = \overline{\hat{f}(y)}$. It follows as in Definition 5.11 that $h(x) = f*g(x)$ is a continuous FV-function whose Fourier transform is given by

$$\hat{h}(y) = \hat{f}(y)\hat{g}(y) = |\hat{f}(y)|^2 \geqslant 0.$$

It follows from Corollary 5.18 that $h(0) = \int_{-\infty}^{\infty}\hat{h}(y)\,dy$, or that

$$\int_{-\infty}^{\infty}|f(x)|^2\,dx = \int_{-\infty}^{\infty}f(x)g(-x)\,dx = h(0) = \int_{-\infty}^{\infty}|\hat{f}(y)|^2\,dy,$$

as required.

It is worth noting that Theorem 5.20 remains true for much wider classes of functions – for instance, for any f for which both f and f^2 are FV-functions, or even for general f in the class L^2 of (Lebesgue) square-integrable functions. These results can be easily deduced from Theorem 5.20 by a density argument using L^2-norms.

A modification of the proof of Theorem 5.20 (Bessel's equation) gives us another famous identify due to Parseval. Suppose that f, g are any bounded FV-functions. We have seen in Theorem 5.20 that both $\hat{f}^2$ and $\hat{g}^2$ are FV-functions, and it follows from the Cauchy–Schwarz inequality that $\hat{f}\hat{g}$ is also an FV-function.

It follows that $h(x) = f*g(x)$ is a continuous FV-function (whose Fourier transform is $\hat{f}\hat{g}$) which satisfies the hypotheses of Theorem 5.17, and hence that

$$h(x) = \int_{-\infty}^{\infty} \hat{h}(y)e^{2\pi ixy}\,dy \qquad \text{for all real } x.$$

In terms of f and g, this states that

$$\int_{-\infty}^{\infty} f(t)g(x-t)\,dt = \int_{-\infty}^{\infty} \hat{f}(y)\hat{g}(y)e^{2\pi ixy}\,dy$$

or on putting $x = 0$, and replacing g by $g_1(t) = \overline{g(-t)}$ that

$$\int_{-\infty}^{\infty} f(t)\overline{g_1(t)}\,dt = \int_{-\infty}^{\infty} \hat{f}(y)\overline{\hat{g}_1(y)}\,dy.$$

We have just proved the following result.

Theorem 5.21 Let f, g be bounded FV-functions. Then $\hat{f}\overline{\hat{g}}$ is an FV-function and $\int_{-\infty}^{\infty} f(t)\overline{g(t)}\,dt = \int_{-\infty}^{\infty} \hat{f}(y)\overline{\hat{g}(y)}\,dy$. In terms of the inner product defined by $(f,g) = \int_{-\infty}^{\infty} f(t)\overline{g(t)}\,dt$, we have shown that $(f,g) = (\hat{f},\hat{g})$, or that the inner product of two functions is preserved under Fourier transforms.

The results are made clearer by examples, as follows.

Examples 5.22 (i) Let $f(x) = 1$ on $[-a, a]$ for some real $a > 0$ (a special case of Example 5.9(ii)) and zero elsewhere.
Then

$$\hat{f}(y) = \int_{-a}^{a} e^{-2\pi ixy}\,dx = -\frac{1}{2\pi iy}(e^{-2\pi iay} - e^{2\pi iay})$$

$$= \frac{1}{\pi y}\sin(2\pi ay) \qquad \text{if } y \neq 0,$$

and $$\hat{f}(0) = 2a.$$

As already noted under Example 5.9(ii), $\hat{f}$ is not an FV-function: however, f is bounded and as predicted by Theorem 5.20, $\hat{f}^2$ is an FV-function, and we obtain from Bessel's equation that

$$2a = \int_{-\infty}^{\infty} |f(x)|^2\,dx$$

$$= \int_{-\infty}^{\infty} |\hat{f}(y)|^2\,dy$$

$$= \frac{1}{\pi^2}\int_{-\infty}^{\infty} \frac{\sin^2(2\pi ay)}{y^2}\,dy,$$

a result which is proved in a different way in Theorem B.5(i).
(ii) A similar calculation starting from

$$f(x) = \begin{cases} 1 & \text{on } (0, a), \\ -1 & \text{on } (-a, 0) \end{cases}$$

gives

$$\hat{f}(y) = 2i\frac{\sin^2(\pi a y)}{\pi y} \qquad (\text{and } \hat{f}(0) = 0).$$

Here again, $\hat{f}$ is not an FV-function: this time the limit $\int_0^A \hat{f}(y)\,dy \to \infty$ as $A \to \infty$, but Bessel's equation is still valid and gives a result equivalent to that in (i).

5.5 THE DIRICHLET PROBLEM IN THE UPPER HALF-PLANE, AND THE HILBERT TRANSFORM

The problem of finding a harmonic function in a disc, which has prescribed boundary values, was studied in Chapter 3 using the Fourier coefficients of the boundary function to construct the required solution. The analogous problem when considering Fourier integrals is to regard the real line as embedded in the plane $\mathbb{R}^2$ and being the boundary of one of the resulting half-planes – conventionally the upper half-plane, which we shall denote by $H = \{(x, y): y > 0\}$. We suppose that a function f is given on the line and that a harmonic function is required on H which will have f as its boundary values: this will be the appropriate version of the Dirichlet problem in the present context.

One difference between the two settings is immediately apparent. Since the upper half-plane is an unbounded set, it is possible for a harmonic function to vanish on the line, but not in H: indeed, if we take *any* function which is analytic in the whole plane and real valued on the real axis, then its imaginary part has the required property. Suitable examples are given by $\text{Im}(z^n)$, $n = 1, 2, 3, \ldots$; $\text{Im}(e^z)$; $\text{Im}(\cos z)$; etc. Hence any such function may be added to a solution of the boundary value problem to get further solutions, and to obtain a unique solution we have to impose a condition on its behaviour at infinity – usually that it should be bounded

It would be possible, though rather lengthy, to obtain the solution of the Dirichlet problem along lines parallel to the development in Chapter 3, but without making direct use of those results, and it can be argued that this self-contained approach is the most satisfactory. However, we observe that the bilinear mapping

$$\phi(x) = \frac{1 + ix}{1 - ix}$$

takes the real axis onto the unit circle (with -1 omitted), the upper half-plane onto the interior of the circle (since $\phi(i) = 0$), and recall (Theorem 3.19) that if h is harmonic and ϕ is analytic then $h \circ \phi$ is again harmonic. Thus solutions to Dirichlet problems in one or other contexts are mutually interchangeable via the mapping ϕ, and results in one can be deduced at once from those in the other. This approach, which we shall follow in this section, makes up in economy what it lacks in self-containedness. It has the added advantage of largely avoiding the problem of non-uniqueness described above.

We begin with an informal derivation of the form of the Poisson integral which is appropriate here, and use it to derive some examples, before beginning any formal deductions. Since this is an informal derivation we shall not be too fussy about the hypotheses satisfied by the functions under consideration, though it will become clear what these ought to be, and we shall point this out at the end of the calculation.

Suppose then that f is a real- (or complex-) valued function defined on $\mathbb{R}$, and we require a function h which is defined and harmonic in the upper half-plane and has f for its boundary values: $h(x, y) \to f(x)$ as $y \to 0_+$. We shall use ϕ for the bilinear mapping

$$\phi(x) = \frac{1 + ix}{1 - ix}$$

considered above; so that $f_1 = f \circ \phi^{-1}$ is defined on the unit circle.

It follows that the Poisson integral Pf_1 of f_1, given by

$$Pf_1(re^{it}) = (f_1 * P_r)(t)$$

$$= \frac{1}{2\pi} \int_0^{2\pi} f_1(e^{iu}) P_r(t - u) \, du$$

where P_r is the Poisson kernel given by Definition 3.3, is the required solution to the corresponding Dirichlet problem on the disc.

Since

$$P_r(t) = \text{Re}\left(\frac{1 + re^{it}}{1 - re^{it}}\right)$$

we can write the above as

$$Pf_1(re^{it}) = \frac{1}{2\pi} \int_0^{2\pi} f_1(e^{iu}) \text{Re}\left(\frac{1 + re^{i(t-u)}}{1 - re^{i(t-u)}}\right) du$$

$$= \frac{1}{2\pi} \int_0^{2\pi} f \circ \phi^{-1}(e^{iu}) \text{Re}\left(\frac{e^{iu} + re^{it}}{e^{iu} - re^{it}}\right) du,$$

or as

$$Pf_1(z) = \frac{1}{2\pi i} \int_c f \circ \phi^{-1}(w) \text{Re}\left(\frac{w + z}{w - z}\right) \frac{dw}{w}$$

where we have put $z = re^{it}$ and $w = e^{iu}$, and C denotes the unit circle, $|w| = 1$.

But we can now put $w = \phi(x)$ for real x, so that $f \circ \phi^{-1}(w) = f(x)$, and in addition put $z = \phi(v)$ where $v = x + iy (y > 0)$ is in the upper half-plane. We then have

$$w = \frac{1 + ix}{1 - ix} \qquad \text{and so}$$

$$\frac{dw}{dx} = \frac{i(1 - ix) + i(1 + ix)}{(1 - ix)^2} = \frac{2i}{(1 - ix)^2}$$

or

$$\frac{dw}{w} = \frac{2i\,dx}{1+x^2}.$$

The expression

$$\frac{w+z}{w-z} \quad \text{becomes}$$

$$\frac{\phi(x)+\phi(v)}{\phi(x)-\phi(v)} = \frac{1}{i}\left(\frac{1+xv}{x-v}\right)$$

as is immediately verified. The above integral then becomes

$$Pf_1(\phi(v)) = \frac{1}{\pi}\int_{-\infty}^{\infty} f(x)\,\mathrm{Re}\left\{\frac{1}{i}\left(\frac{1+xv}{x-v}\right)\right\}\frac{dx}{1+x^2},$$

and we finish the calculation by finding a more amenable form for the kernel: to do this we put t for x, and $v = x + iy$.
Then

$$\mathrm{Re}\left\{\frac{1}{i}\frac{1+tv}{t-v}\right\} = \mathrm{Re}\left\{\frac{1}{i}\frac{1+tx+ity}{(t-x-iy)}\right\} = \mathrm{Re}\left\{\frac{(1+tx+ity)(t-x+iy)}{i((t-x)^2+y^2)}\right\}$$

$$= \frac{ty(t-x)+y(1+tx)}{(t-x)^2+y^2} = \frac{y(1+t^2)}{(t-x)^2+y^2}.$$

This leads us to the following expression for $F(v) = Pf_1 \circ \phi(v)$:

$$F(v) = \frac{1}{\pi}\int_{-\infty}^{\infty} f(t)\frac{y}{(t-x)^2+y^2}\,dt$$

$$= (f * K_y)(x), \qquad \text{where } K_y(x) = \frac{y}{\pi(x^2+y^2)}$$

is the appropriate Poisson kernel for the half-plane.

Thus we have found that the solution to the Dirichlet problem for f is given by $f * K_y$. The condition which f must satisfy is that $f_1 = f \circ \phi^{-1}$ should be an FC-function on the unit circle, i.e. it should be bounded and have only a finite number of discontinuities. In terms of f itself this means that it must be bounded on $\mathbb{R}$, have only a finite number of discontinuities, and have finite limits as $x \to \pm\infty$. These are not quite the conditions that f should be an FV-function as defined in Section 5.2, but the difference is not crucial.

We illustrate all this with some simple examples.

Examples 5.25 (i) Let $f(t) = 1/(1+t^2)$ for real t.
We follow through the above procedure, beginning by putting

$$z = \phi(t) = \frac{1+it}{1-it} \qquad \text{and thus}$$

$$f_1(z) = f \circ \phi^{-1}(z) = \left\{ 1 - \left(\frac{1-z}{1+z} \right)^2 \right\}^{-1} = \frac{1}{2} + \frac{1+z^2}{4z} = \frac{1}{2} + \frac{1}{4}(z + \frac{1}{2}).$$

If we put $z = e^{i\theta}$, we get $f_1(z) = \frac{1}{2}(1 + \cos\theta)$, and so the Poisson integral

$$Pf_1(re^{i\theta}) = \frac{1}{2}(1 + r\cos\theta)$$
$$= \frac{1}{2}(1 + \operatorname{Re} z)$$

where $z = re^{i\theta}$ is now regarded as a point of the unit disc, $|z| \leqslant 1$.

Hence if we put $z = \phi(v)$ where $v = x + iy, y > 0$, then we obtain the Poisson integral of f in the form

$$F(v) = \frac{1}{2} \left\{ 1 + \operatorname{Re}\left(\frac{1 + i(x+iy)}{1 - i(x+iy)} \right) \right\}$$
$$= \frac{1}{2} \left\{ 1 + \operatorname{Re}\left(\frac{1 - y + ix}{1 + y - ix} \right) \right\}$$
$$= \frac{1}{2} \left\{ 1 + \frac{1 - y^2 - x^2}{(1+y)^2 + x^2} \right\} = \frac{1+y}{(1+y)^2 + x^2}.$$

This obviously reduces to $1/(1 + x^2)$ as required when y is zero, and is bonded for $y \geqslant 0$.

This is an example in which a direct application of the convolution $f * K_y$ would involve a tedious calculation with partial fractions. Our next example shows the opposite situation.

(ii) Let

$$f(t) = \begin{cases} 1 & \text{for } t \geqslant 0, \\ 0 & \text{for } t < 0 \end{cases}$$

Then the Poisson integral $F(v) = F(x + iy)$ is given by

$$\frac{1}{\pi} \int_{-\infty}^{\infty} f(t) K_y(x - t) \, dt = \frac{1}{\pi} \int_0^{\infty} \frac{y \, dt}{(x-t)^2 + y^2}$$
$$= \frac{1}{\pi} \tan^{-1}\left(\frac{t-x}{y} \right) \bigg|_0^{\infty}$$
$$= \frac{1}{\pi} \left(\frac{\pi}{2} + \tan^{-1}\frac{x}{y} \right).$$

If we put $x + iy = re^{i\theta}$ with $0 < \theta < \pi$ for $y > 0$, then we have

$$\theta = \frac{\pi}{2} - \tan^{-1}\frac{x}{y} \quad \text{and hence} \quad F(v) = \frac{1}{\pi}(\pi - \theta) = 1 - \frac{\theta}{\pi}.$$

This result can also be seen by inspection since $\theta = \operatorname{Im}(\log z)$ is harmonic and is equal to zero for $x > 0, y = 0$, and to π for $x < 0, y = 0$. Notice also that f is not an FV-function since $\int_0^{\infty} f(t) \, dt$ does not exist, but that $f \circ \phi^{-1}$ is an FC-function on the circle, as mentioned above.

To begin the more formal study of the Poisson integral in the half-plane we define the Poisson kernel with its elementary properties, and then proceed to the principal theorem of the section.

Definition 5.24   Let $K_y(x) = y/(\pi(x^2 + y^2))$ for real x, and $y > 0$. Then we have

(i) $K_y(x) > 0$;

(ii) $\displaystyle \int_{-\infty}^{\infty} K_y(x)\,dx = 1$;

(iii) for $\delta > 0$, $\displaystyle \int_{-\infty}^{-\delta} K_y(x)\,dx + \int_{\delta}^{\infty} K_y(x)\,dx \to 0$ as $y \to 0$.

K_y is called the *Poisson kernel* for the half-plane, and the properties (i), (ii), and (iii) are immediate since K can be integrated in terms of the inverse tangent.

Definition 5.25   Let f be an FV-function on the real line. Then the function $F(x + iy) = (f * K_y)(x)$

$$= \frac{1}{\pi} \int_{-\infty}^{\infty} f(t) \frac{y}{(t-x)^2 + y^2}\,dt$$

is called the *Poisson integral* of f. The notation $F = Pf$ will also be used when there is no danger of confusion with the Poisson integral on the disc. It has the following properties, which are analogous to those in Theorem 3.6.

Theorem 5.26   Let f be a bounded FV-function and $F(x + iy) = (f * K_y)(x)$ be its Poisson integral as defined above, for $y > 0$.
   Then F is a bounded harmonic function for $y > 0$, and

$$F(x + iy) \to f(x) \qquad \text{as } y \to 0_+$$

at each point of continuity of f.

Proof   If we suppose that $|f(x)| \leqslant M$ for all real x, then

$$|F(x + iy)| \leqslant \frac{M}{\pi} \int_{-\infty}^{\infty} \frac{y}{(x-t)^2 + y^2}\,dt = M,$$

so that F has the same bounds as f does. Similarly, since

$$K_y(x) = \frac{y}{\pi(x^2 + y^2)} = -\operatorname{Im}\left(\frac{1}{\pi z}\right)$$

is harmonic, and for $y > 0$ we may differentiate the integral defining F under the integral sign, it follows that F must be harmonic also.
   To show that $F(x + iy) \to f(x)$ at a point of continuity of f, we let x be such a point, and given $\varepsilon > 0$, choose $\delta > 0$ so that

$$|f(x) - f(t)| < \varepsilon \qquad \text{when } |x - t| < \delta.$$

Using property (ii) of Definition 5.24, we can write

$$F(x + iy) - f(x) = \frac{1}{\pi} \int_{-\infty}^{\infty} (f(t) - f(x))K_y(x - t)\,dt,$$

and we divide the range of integration into $(-\infty, x - \delta], (x - \delta, x + \delta), [x + \delta, \infty)$.
It follows that

$$\left| \int_{x-\delta}^{x+\delta} (f(t) - f(x))K_y(x - t)\,dt \right|$$

$$\leqslant \varepsilon \int_{-\infty}^{\infty} K_y(x - t)\,dt = \varepsilon, \qquad \text{using property (i),}$$

while

$$\left| \int_{|t-x|\geqslant\delta} (f(t) - f(x))K_y(x - t)\,dt \right|$$

$$\leqslant 2M \int_{|x-t|\geqslant\delta} K_y(x - t)\,dt$$

which tends to zero with y by property (iii), and the result follows.

When we turn to investigate the function which is the harmonic conjugate in H of Pf, we come across the same sort of problems which occurred in Chapter 4. We begin as usual with an informal derivation of the required formulae.
Observe that the Poisson kernel

$$K_y(x) = \frac{y}{\pi(x^2 + y^2)}$$

can be written in the form

$$\text{Re}\left\{ \frac{i(x - iy)}{\pi(x^2 + y^2)} \right\} = \text{Re}\left\{ \frac{i}{\pi(x + iy)} \right\}.$$

Its harmonic conjugate $\tilde{K}_y(x)$ is therefore given by

$$\text{Im}\left(\frac{1}{\pi z} \right) = \text{Im}\left(\frac{i(x - iy)}{\pi(x^2 + y^2)} \right) = \frac{x}{\pi(x^2 + y^2)},$$

and we find that harmonic conjugate of $Pf(x)$ is given by

$$\tilde{P}f(x) = f * \tilde{K}_y(x) = \frac{1}{\pi} \int_{-\infty}^{\infty} f(t)\frac{(x - t)}{(x - t)^2 + y^2}\,dt = \frac{1}{\pi} \int_{-\infty}^{\infty} f(x - t)\frac{t}{t^2 + y^2}\,dt.$$

In particular, when $y \to 0$ we obtain formally that the conjugate function of f on the real line is given by

$$\frac{1}{\pi} \int_{-\infty}^{\infty} \frac{f(t)}{x - t}\,dt = \frac{1}{\pi} \int_{-\infty}^{\infty} \frac{f(x - t)}{t}\,dt$$

$$= \frac{1}{\pi} \int_{0}^{\infty} \{f(x - t) - f(x + t)\}\frac{dt}{t}.$$

After our experience in Chapter 4, we should expect to have to interpret this integral in one or other of the forms

$$\lim_{\delta\to 0_+} \frac{1}{\pi}\int_{|x-t|\geqslant\delta}\frac{f(t)}{x-t}\,\mathrm{d}t \quad \text{or} \quad \lim_{\delta\to 0_+}\frac{1}{\pi}\int_\delta^\infty \{f(x-t)-f(x+t)\}\frac{\mathrm{d}t}{t}.$$

This leads us to the following definitions.

Definition 5.27 (i) Let f be an FV-function on the real line. The function

$$\tilde{K}_y(x) = \frac{x}{\pi(x^2+y^2)} \qquad (y>0)$$

is called the *conjugate Poisson kernel*, and $\tilde{P}f(z) = (f*\tilde{K}_y)(x)$, $z = x+\mathrm{i}y$, is called the *conjugate Poisson integral* of f.

(ii) The integral

$$\frac{1}{\pi}\int_{-\infty}^\infty \frac{f(t)}{x-t} = \frac{1}{\pi}\int_0^\infty \{f(x-t)-f(x+t)\}\frac{\mathrm{d}t}{t}$$

(defined as above by a limit when this exists) is called the *Hilbert transform* of f and denoted by $Hf(x)$.

General questions concerning the existence and properties of Hf are beyond our scope: we shall be content with the elementary observations in the following result,

Theorem 5.28 Let f be an FV-function. Then Hf exists at every point x at which there is a constant M for which

$$|f(x+h)-f(x-h)| \leqslant Mh$$

for small h (say $0<h<1$).

In particular, Hf exists at every point at which f is Lipschitz continuous. Hf is continuous on every interval I on which the above condition holds (for a fixed value of M) whenever the points $x\pm h$ are in I.

Proof Let f be any FV-function. We begin by showing that the integral

$$\int_\delta^\infty \frac{f(x)}{x}\,\mathrm{d}x$$

exists, for any $\delta>0$. For let $F(x)=\int_\delta^x f(t)\,\mathrm{d}t$. Since f is an FV-function, F is a continuous function with $F(\delta)=0$ and a finite limit as $x\to\infty$.

It follows on integrating by parts that

$$\int_\delta^\alpha \frac{f(x)}{x}\,\mathrm{d}x = \left[\frac{F(x)}{x}\right]_\delta^\alpha + \int_\delta^\alpha \frac{F(x)}{x^2}\,\mathrm{d}x,$$

and if we let $\alpha \to \infty$ the right-hand side approaches the finite value

$$\int_\delta^\infty \frac{F(x)}{x^2} \, dx$$

as required.

It follows that in the integral defining $Hf(x)$, namely

$$\lim_{\delta \to 0_+} \int_\delta^\infty \{f(x-t) - f(x+t)\} \frac{dt}{t} = \lim_{\delta \to 0_+} \left(\int_\delta^1 + \int_1^\infty \right) \{f(x-t) - f(x+t)\} \frac{dt}{t},$$

the integral over $[1, \infty)$ exists for all FV-functions, and we have only to consider the limit of the integral over $[\delta, 1]$. But the given condition on f shows that the integrand is bounded near $t = 0$ and is thus an FC-function. Consequently the integral over $[0, 1]$ exists, which completes the existence proof. The proof of continuity of Hf is similar to those in Theorems 4.7 and 2.5, and is omitted.

Examples of the Hilbert transform of particular functions can be found easily in a few cases. For an extended list the reader can consult Erdelyi et al. (1954). The following are parallel to those in Example 5.23.

Example 5.29   (i) Let $f(t) = 1/(1 + t^2)$ for real t.
Then as noted in Example 5.23(i),

$$Pf(z) = \frac{1}{2}\left(1 + \text{Re}\left(\frac{1 + iz}{1 - iz} \right) \right)$$

and consequently its harmonic conjugate

$$\tilde{P}f(z) = \frac{1}{2}\text{Im}\left(\frac{1 + iz}{1 - iz} \right).$$

Putting $x = x + iy$, this gives

$$\tilde{P}f(z) = \frac{1}{2}\text{Im}\left\{ \frac{(1 + ix - y)(1 + ix + y)}{(1 - ix + y)(1 + ix + y)} \right\}$$

$$= \frac{1}{2}\frac{x(1 + y) + x(1 - y)}{x^2 + (1 + y)^2} = \frac{x}{x^2 + (1 + y)^2}.$$

Hence putting $y = 0$ and t for x, we obtain

$$Hf(t) = \frac{t}{1 + t^2}$$

This calculation of course does no more than verify that $\tilde{K}_y$ is conjugate to K_y.

(ii) Let

$$f(t) = \begin{cases} 1 & \text{for } t \geqslant 0 \\ 0 & \text{for } t < 0. \end{cases}$$

We showed in Example 5.23 (i) that the Poisson integral

$$Pf(z) = 1 - \frac{1}{\pi} \operatorname{Im}(\log z).$$

and hence that its harmonic conjugate

$$\tilde{P}f(x) = \frac{1}{\pi} \operatorname{Re}(\log z) = \frac{1}{\pi} \log |z|.$$

Hence for real t, $Hf(t) = (1/\pi)\log|t|$. Direct evaluation of the convolution integral $f * \tilde{K}_y$ is not possible here since f is not an FV-function.

5.6 SOME APPLICATIONS OF THE FOURIER INTEGRAL

As we said in the Preface, the domain of applications of Fourier theory extends into nearly every area of scientific and technological study, and no attempt at any kind of completeness can be attempted here. Instead we have chosen a small number of applications which are either interesting in themselves, or illustrate a part of the previous theory, or bring to light a further piece of non-trivial mathematical reasoning (or hopefully all three).

We first derive the well-known result on the vibration of an infinitely long elastic string.

The motion of such a string is governed by the partial differential equation

$$\frac{\partial^2 f}{\partial t^2} = c^2 \frac{\partial^2 f}{\partial x^2}$$

for the displacement f, where c^2 denotes a positive quantity depending on the physical properties (density and elasticity) of the string, and will be constant for a uniform string. The solution will be subject to suitable initial and boundary conditions. Since we are considering an infinite string, the appropriate boundary condition is that the solution should remain bounded for large x (or equivalently, as it turns out, for large t). For an initial condition we shall suppose that the string is given a displacement $f(x, 0) = f_0(x)$ and is released from this position from rest, i.e. that $(\partial/\partial t)f(x, 0) = 0$ for all x. (For completeness we might as well add that physical constraints on the nature of the string prevent f_0 from being of too pathological a nature – we could reasonably suppose that it is zero except on some finite interval on which it is Lipschitz continuous, for example.)

We begin as we did in the example at the start of Chapter 1 by looking for solutions of the form $f(x, t) = g(x)h(t)$ in which the variables are separated.

Substituting this in the original equation gives

$$g(x)h''(t) = c^2 g''(x)h(t),$$

$$\text{or} \quad \frac{h''(t)}{h(t)} = c^2 \frac{g''(x)}{g(x)},$$

and since x and t can vary independently it follows that both sides of the last equation must be constant.

A bounded solution of the equation $g''(x)/g(x) = \text{constant}$ must have the form $g(x) = e^{iyx}$ for some real value of y, and this then gives $h(t) = e^{\pm icyt}$.

Solutions of the original equation can thus be found by superimposing solutions of the form $e^{i(x \pm ct)y}$ where y is an unrestricted real variable. Consequently the solution takes the form

$$f(x, t) = \int_{-\infty}^{\infty} \phi_1(y)e^{i(x + ct)y}\,\mathrm{d}y + \int_{-\infty}^{\infty} \phi_2(y)e^{i(x - ct)y}\,\mathrm{d}y,$$

where the functions, ϕ_1 and ϕ_2 must be found from the initial conditions. These require that

$$f_0(x) = f(x, 0) = \int_{-\infty}^{\infty} \{\phi_1(y) + \phi_2(y)\}e^{ixy}\,\mathrm{d}y,$$

and that

$$0 = \frac{\partial}{\partial t} f(x, 0) = \int_{-\infty}^{\infty} icy\{\phi_1(y) - \phi_2(y)\}e^{ixy}\,\mathrm{d}y.$$

If we apply the inversion theorem (Theorem 5.17) to the second of these equations, we see that $\phi_1(y) = \phi_2(y)$, while from the first we have that

$$\phi_1(y) + \phi_2(y) = \frac{1}{2\pi}\hat{f}_0\left(\frac{y}{2\pi}\right).$$

This gives us the final result in the form

$$f(x, t) = \frac{1}{2}\int_{-\infty}^{\infty}\frac{1}{2\pi}\hat{f}_0\left(\frac{y}{2\pi}\right)e^{i(x + ct)y}\,\mathrm{d}y + \frac{1}{2}\int_{-\infty}^{\infty}\frac{1}{2\pi}\hat{f}_0\left(\frac{y}{2\pi}\right)e^{i(x - ct)y}\,\mathrm{d}y$$

and it remains only to re-apply Theorem 5.17 to obtain

$$f(x, t) = \tfrac{1}{2}f_0(x + ct) + \tfrac{1}{2}f_0(x - ct).$$

This brings us to the well-known result that an initial displacement of an infinite string, if released from rest, breaks into two equal displacements with half the amplitude which travel in opposite directions with constant speed c. The result also holds for finite strings for the initial period until the waves reach the ends, after which end-effects occur.

Our second example recalls even more closely the example on the conduction of heat with which we began Chapter 1. We shall consider the problem of the distribution of heat in an infinite rod, with an initial value given by $f_0(x)$.

The temperature is given by a function $f(x, t)$ where x is the displacement along the rod and t is the (positive) time elapsed from the initial state. The variation in temperature is governed by the equation.

$$\frac{\partial^2 f}{\partial x^2} = k^2 \frac{\partial f}{\partial t}$$

where k^2 is a positive quantity depending on the physical properties of the rod: for simplicity we shall suppose the rod uniform so that k is a constant.

Using the now familiar method of separation of variables, we look for solutions of the form $f(x, t) = g(x)h(t)$, from which we see that

$$\frac{g''(x)}{g(x)} = k^2 \frac{h'(t)}{h(t)}.$$

Both sides of this equation are constant, so that we can take $h(t) = e^{-y^2 t}$, the form of the exponential being determined by the requirement that the temperature will tend to zero with increasing time, and thus y can take any real value. We then obtain $g''(x) + k^2 y^2 g(x) = 0$ from which follows

$$g(x) = \phi(y)e^{ikyx},$$

where in this solution y may be positive or negative.

It follows that a general solution of the differential equation takes the form

$$f(x, t) = \int_{-\infty}^{\infty} \phi(y)e^{ikyx - y^2 t}\, dy,$$

where we assume that ϕ (which we shall shortly determine from f_0) is sufficiently well behaved for the integral to converge for all $t \geqslant 0$: for instance, it would suffice if ϕ were an FV-function.

To make our solution fit the initial conditions, we require $f_0(x) = f(x, 0) = \int_{-\infty}^{\infty} \phi(y)e^{ikyx}\, dy$, from which it follows on putting $ky = 2\pi u$ that

$$\frac{k}{2\pi}f_0(x)$$

is the inverse Fourier transform of $\phi(2\pi u/k)$.
When we invert this relationship we see that

$$\phi\left(\frac{2\pi u}{k}\right) = \int_{-\infty}^{\infty} \frac{k}{2\pi}f_0(x)e^{-2\pi ixu}\, dx, \text{ or that}$$

$$\phi(y) = \frac{k}{2\pi}\int_{-\infty}^{\infty} f_0(x)e^{-ikxy}\, dx,$$

and thus we have determined ϕ in terms of the Fourier transform of f_0. Sufficient conditions for the validity of this procedure are thus that f_0 should satisfy the hypotheses of the Fourier inversion theorem (Theorem 5.17).

If we now substitute this expression for ϕ into the solution for $f(x, t)$ derived above we obtain

$$f(x, t) = \int_{-\infty}^{\infty} \phi(y)e^{ikxy - ty^2}\, dy$$

$$= \frac{k}{2\pi}\int_{-\infty}^{\infty} \left(\int_{-\infty}^{\infty} f_0(u)e^{-ikuy}\, du\right)e^{ikxy - ty^2}\, dy$$

$$= \frac{k}{2\pi} \int_{-\infty}^{\infty} f_0(u) \left(\int_{-\infty}^{\infty} e^{iky(x-u)-ty^2} \, dy \right) du.$$

We can evaluate the inner integral explicitly, using Theorem B.4, to obtain

$$\int_{-\infty}^{\infty} e^{izy-ty^2} \, dy = \int_{-\infty}^{\infty} e^{-z^2/4t} e^{-t(y-iz/2t)^2} \, dy = \sqrt{\frac{\pi}{t}} e^{-z^2/4t}.$$

Hence

$$f(x,t) = \frac{k}{2\pi} \int_{-\infty}^{\infty} f_0(u) \sqrt{\frac{\pi}{t}} e^{-k^2(x-u)^2/4t} \, du,$$

and we have obtained the solution in the form of a convolution in which the Fourier transform no longer appears explicitly:

$$f(x,t) = \frac{k}{2\pi} f_0 * g_t(x), \qquad \text{where } g_t(x) = \sqrt{\frac{\pi}{t}} e^{-k^2x^2/4t}.$$

A case in which we can find $f(x,t)$ explicitly occurs when $f_0(x)$ is already of the form e^{-cx^2}. We then get

$$f(x,t) = \frac{k}{2\pi} \int_{-\infty}^{\infty} e^{-cu^2} \sqrt{\frac{\pi}{t}} e^{-k^2(x-u)^2/4t} \, du,$$

and if we apply Theorem B.4 as above, after a little reduction we obtain

$$f(x,t) = \frac{1}{\sqrt{(1+vt)}} e^{-cx^2/(1+vt)} \qquad \text{where } v = \frac{4c}{k^2}.$$

Our final application of the Fourier integral concerns a summation formula due to Poisson, which is often useful in the summation of series. As usual we first derive the formula informally, and then look for hypotheses under which it is valid.

Suppose initially that f is a continuous function on $\mathbb{R}$, and consider the function $g(x) = \sum_{n=-\infty}^{\infty} f(x+n)$, where we suppose that f tends to zero sufficiently rapidly for the series to converge: this is by no means the case for all FV-functions, as several of the examples in this chapter show. However a bound of the form

$$|f(x)| < \frac{K}{1+x^2}$$

is clearly sufficient to make the series absolutely and uniformly convergent and it then follows that g too will be continuous.

We now observe that g is of period 1 and hence may be expanded by a Fourier series on $[0,1]$: say $S(g) = \sum_{-\infty}^{\infty} c_n(g)e^{2\pi inx}$ where $c_n(g) = \int_0^1 g(x)e^{-2\pi inx} \, dx$. If we suppose further that $S(g)$ is uniformly convergent, then by our first result on the convergence of Fourier series in Theorem 1.11, the series converges to g:

$$g(x) = \sum_{-\infty}^{\infty} c_n(g)e^{2\pi inx}.$$

The interest of this formula lies in the form it takes when one puts g and $c_n(g)$ in terms of the original function f. In particular, we obtain

$$c_n(g) = \int_0^1 g(x)e^{-2\pi inx}\,dx = \int_0^1 \sum_{m=-\infty}^{\infty} f(x+m)e^{-2\pi inx}\,dx$$

$$= \sum_{m=-\infty}^{\infty} \int_0^1 f(x+m)e^{-2\pi inx}\,dx$$

(assuming as before the uniform convergence of the series)

$$= \sum_{m=-\infty}^{\infty} \int_m^{m+1} f(x)e^{-2\pi inx}\,dx \qquad \text{(putting } x \text{ for } x+m)$$

$$= \int_{-\infty}^{\infty} f(x)e^{-2\pi inx}\,dx$$

$$= \hat{f}(n).$$

We have obtained the surprising result that the Fourier coefficients (c_n) of g are the values of the Fourier transform of f at integer values. Consequently the above formula becomes

$$\sum_{-\infty}^{\infty} f(x+m) = \sum_{-\infty}^{\infty} \hat{f}(n)e^{-2\pi inx} \qquad \text{or simply}$$

$$\sum_{-\infty}^{\infty} f(m) = \sum_{-\infty}^{\infty} \hat{f}(n) \qquad \text{when } x = 0.$$

A further pleasant feature of this situation is that when applying the formula, the typical situation is one in which both f and $\hat{f}$ are known in advance, so that there is no loss in utility in putting hypotheses on both f and $\hat{f}$, something which we have previously avoided whenever possible.

If we now re-examine the above argument we see that apart from the restriction on f already mentioned, it is sufficient for $\sum c_n(g) = \sum \hat{f}(n)$ to be absolutely convergent. We have thus proved the following result.

Theorem 5.30 Let f be an FV-function with the additional properties (i) $|f(x)| < K/(1+x^2)$ for all real x, and (ii) $\sum_{-\infty}^{\infty} \hat{f}(n)$ absolutely convergent. Then for all real x, $\sum_{-\infty}^{\infty} f(x+m) = \sum_{-\infty}^{\infty} \hat{f}(n)e^{-2\pi inx}$, both series being absolutely convergent.

To illustrate this result we give the following examples (see also exercise 16 at the end of the chapter). We begin with Jacobi's theta function identity.

Exercise 5.31 For $c > 0$, $\sum_{m=-\infty}^{\infty} e^{-c(x+m)^2} = \sqrt{\pi/c}\,\sum_{n=-\infty}^{\infty} e^{-\pi^2 n^2/c - 2\pi inx}$. In particular, putting $x = 0, \frac{1}{2}$ we obtain

$$\sum_{-\infty}^{\infty} e^{-cm^2} = \sqrt{\frac{\pi}{c}}\,\sum_{-\infty}^{\infty} e^{-\pi^2 n^2/c} \qquad \text{and}$$

$$\sum_{-\infty}^{\infty} e^{-c(m+1/2)^2} = \sqrt{\frac{\pi}{c}} \sum_{-\infty}^{\infty} (-1)^n e^{-\pi^2 n^2/c} \text{ respectively.}$$

This result is immediate from the pair $f(x) = e^{-cx^2}$, $\hat{f}(y) = \sqrt{\pi/c}\, e^{-\pi^2 g^2/c}$.

Example 5.32 If $f(x) = \begin{cases} \cos \pi x & \text{on } [-\frac{1}{2}, \frac{1}{2}], \\ 0 & \text{elsewhere} \end{cases}$

then it is easy to verify that

$$\hat{f}(y) = \frac{2}{\pi} \frac{\cos(\pi y)}{1 - 4y^2}.$$

The Poisson formula then gives

$$\sum_{-\infty}^{\infty} f(m) = f(0) = 1 = \sum_{-\infty}^{\infty} \hat{f}(n) = \frac{2}{\pi} \sum_{-\infty}^{\infty} \frac{(-1)^n}{1 - 4n^2}$$

$$= \frac{2}{\pi} \left[1 + 2 \sum_{n=1}^{\infty} \frac{(-1)^{n-1}}{4n^2 - 1} \right].$$

EXERCISES

1 Prove that the set Γ of characters of a group $(G, +)$ forms a group under multiplication $(\chi_1 + \chi_2)(x) = \chi_1(x) \cdot \chi_2(x)$ (Lemma 5.4(i))

2 (Odd and even functions.) Show that if f is an even function $(f(x) = f(-x)$ for all real x) then $\hat{f}$ is again even, and both the Fourier transform and its inverse are given by $f \rightarrow 2\int_0^\infty f(x) \cos(2\pi x y)\, dx$. Find a similar result when f is an odd function. (For functions defined initially on $(0, \infty)$ these mappings define the Fourier cosine and sine transforms respectively).

3 Find the Fourier transform of the following functions;

(i) $f(x) = \begin{cases} \cos \dfrac{\pi}{2} x, & |x| \leqslant 1 \\ 0, & |x| > 1 \end{cases}$

(ii) $f(x) = e^{-|x|} \sin x$

(iii) $f(x) = e^{-|x|} \dfrac{\sin x}{x}$

(iv) $f(x) = \dfrac{1}{\cosh ax} \quad (a > 0)$

(*Hint*: use Theorem B.6 in the extended form

$$\int_{-\infty}^{\infty} \frac{e^{zt}}{1 + e^t}\, dt = \frac{\pi}{\sin \pi z} \quad \text{for } 0 < \operatorname{Re} z < 1).$$

4 Let $f(x) = \begin{cases} 1 & \text{for } -1 \leqslant x \leqslant 1 \\ 0 & \text{elsewhere} \end{cases}$

Evaluate $f*f$ and $f*f*f$ for this function.

5 Find the Fourier transform of the function

$$\phi_t(x) = \begin{cases} 1 - t|x| & \text{for } |x| \leqslant t^{-1} \\ 0 & \text{for } |x| > t^{-1} \end{cases}$$

and show that it has the properties (a), (b), (c) listed in Theorem 5.17

6 Repeat exercise 5 for $\phi_t(x) = e^{-tx^2}$.

7 Investigate the summability of $\int_0^\infty \sin mx \, dx$ by Gauss's and Cesàro's methods for values of m which may be complex.

8 Find the Poisson sums of the integrals $\int_0^\infty \cos mx$, $\int_0^\infty x \sin mx$, $\int_0^\infty x \cos mx$, $\int_{-\infty}^\infty \cos mx$.

9 Investigate what modifications need to be made to the proof of Theorem 5.16 if we use the Gauss or Cesàro methods in place of Poisson's. Show that the result is in fact true for *any* function $\phi_t(x)$ which is (a) even; (b) decreasing on $[0, \infty)$ with a piecewise continuous derivative; and (c) has $\phi_t(0) = 1$, $\phi_t(x) \to 1$ as $t \to 0$ for all $x > 0$.

10 Let g be any positive FV-function with $\int_{-\infty}^\infty g(x) dx = 1$. Show that the family of functions defined by $g_t(x) = (1/t)g(x/t)$, $t > 0$, satisfies (a), (b), and (c) of Theorem 5.17.

11 Verify the inversion theorem for the functions in exercise 3 (i), (ii), and (iii). (You may find some of the results in Appendix B helpful.)

12 Modify the proof of Theorem 5.26 to show that if f has right- and left-hand limits l_1 and l_2 at x_0, then $F(x + iy) \to \frac{1}{2}(l_1 + l_2)$ as $y \to 0_+$.

13 Let $f(x) = \begin{cases} 1 & \text{if } 0 < x < a \\ 0 & \text{if } x \leqslant 0 \text{ or } x \geqslant a \end{cases}$

where a is some positive real number. Find Pf, the Poisson integral of f in the upper half-plane, and show that Hf, its Hilbert transform, is given by

$$Hf(x) = \frac{1}{\pi} \log \left| \frac{x-a}{x} \right|. \qquad \text{(Cf. Example 5.23(ii).)}$$

14 Define a sequence of functions $(\phi_n)_{-\infty}^\infty$ by

$$\phi_n(x) = \frac{\sin \pi(x-n)}{\pi(x-n)} \qquad \text{for real } x.$$

Show that this sequence is orthonormal on the real line, i.e. that $\int_{-\infty}^\infty \phi_n(x)\phi_m(x) \, dx = 1$ if $m = n$, and is zero otherwise. Deduce that if (α_n) is a finite sequence of complex numbers ($\alpha_n = 0$ if $|n| \geqslant N$, say) and we define

$$g(x) = \sum_{-N}^N \alpha_n \phi_n(x)$$

then $g(n) = \alpha_n$ for all integers n, and $\int_{-\infty}^\infty g^2(x) \, dx = \sum_{-N}^N \alpha_n^2$.

138

(Advanced readers will consider less restrictive conditions on the coefficient sequence.)

15 Verify that the solution $f(x,t) = (1 + vt)^{-1/2} e^{-cx^2/(1+vt)}$, where $v = 4c/k^2$, obtained in Section 5.6, does indeed satisfy the heat equation

$$\frac{\partial^2 f}{\partial x^2} = k^2 \frac{\partial f}{\partial t} \qquad \text{for all } x \text{ and } t \geqslant 0.$$

16 What previous result is obtained from the Poisson summation formula (5.30) when $f(x) = 1/(a^2 + x^2) \quad (a > 0)$?

CHAPTER 6

Multiple Fourier Series
and Integrals

For our final chapter we shall consider the extension of our theory to functions of several variables. We shall encounter a number of new features in this situation of varying degrees of difficulty. The first is the question of notation, which is most easily handled using vectors and the complex exponential form of the series and integrals. Secondly, there is the technical matter of the convergence of multiple series or integrals. This is discussed in detail in Appendix A: for the present we simply note that when dealing with double series for instance, we shall be concerned with existence of limits of the form $\lim_{m,n \to \infty} \sum_{-m}^{m} \sum_{-n}^{n} f(m, n)$, that is with the existence of *double* (as opposed to *iterated*) limits of symmetric sums in the sense of Definition A.36. The next problem is that of specifying the class of functions to which our theory applies: this will occupy us in Section 6.1, where we shall define appropriate analogues of the FC-functions of Chapter 1.

Finally, there is one new and quite unexpected feature which makes the theory in several variables in some ways quite different from that in a single variable. This is the fact, which is illustrated in Theorem 6.11, that a function may be zero in a neighbourhood of a point, and yet its Fourier series may not converge at that point. It follows from this surprising fact that no purely local condition on f can ensure the convergence of its Fourier series at a point. (From a more sophisticated viewpoint we could say that in order to ensure convergence at a point we must redefine our idea of a neighbourhood of the point, but we shall not pursue this idea here.)

It should be noted that the case of dimension 2 shows all the essential difficulties of the general case, while allowing some notational simplification since we do not have to use suffixes extensively. For this reason we shall generally use the case $n = 2$ to illustrate our ideas.

With this introduction we can begin our systematic study by defining the type of function which we shall use.

6.1 FC- AND FV-FUNCTIONS OF SEVERAL VARIABLES: FOURIER COEFFICIENTS AND SERIES

In Chapter 1 we introduced the class of FC-functions which we defined as those functions which were bounded on $[0, 2\pi]$ and had only a finite number of discontinuities. Let us suppose that f, g are two such functions, and define a function h of two variables by

$$h(x, y) = f(x)g(y).$$

We shall certainly require that such a function h should be integrable: indeed we shall see shortly that its Fourier series is simply the product of the series for f and g. However, discontinuity of f (at x_0 say) produces discontinuities of h along the *line* $x = x_0$. This shows that we must admit functions which are discontinuous not just at isolated points but along lines. In this example the lines are of course parallel to the axes, but it is also convenient to allow lines at other angles. We retain from Chapter 1 the requirement that our functions must be bounded and this results in the following definition.

Definition 6.1 (i) Let f be a real- or complex-valued function on $R = [0, 2\pi] \times [0, 2\pi]$. (After the discussion following Definition 5.2 we can also denote R by $\mathbb{T}^2$.)
We say that f is an *FC function* on R if (and only if)
  (a) f is bounded on R, and
  (b) there is a set A in R, consisting of the union of finite number of line segments (which may or may not be parallel to one or other axis) such that all discontinuities of f are in the set A.
(ii) Let f be a real- or complex-valued function on

$$\mathbb{T}^p = [0, 2\pi] \times [0, 2\pi] \times \cdots \times [0, 2\pi].$$

We say that f is an *FC function* on $\mathbb{T}^p$ if (and only if)
  (a) f is bounded on $\mathbb{T}^p$, and
  (b) there is a set A in $\mathbb{T}^p$ consisting of the union of a finite number of planes $((p-1)$-dimensional linear subspaces) which contains all discontinuities of f.

The idea behind this definition is that it allows discontinuities of f, but the function is not allowed to become large near such a point, and the set of discontinuities is of a restricted type, which none the less is sufficient for many interesting applications. We shall assume in addition that the definition of f is extended by periodicity ($f(x + 2m\pi, y + 2n\pi) = f(x, y)$), since any resulting discontinuities will also be on line segments corresponding to the boundaries of $\mathbb{T}^2$.

It will occasionally be convenient to replace $[0, 2\pi]$ by a translation such as $[-\pi, \pi]$, as is done in the following example.

Example 6.2 (i) For $|x| \leqslant \pi$, $|y| \leqslant \pi$, let

$$f(x, y) = \begin{cases} \dfrac{xy}{x^2 + y^2}, & (x, y) \neq (0, 0) \\ 0, & (x, y) = (0, 0). \end{cases}$$

Extend f to $\mathbb{R}^2$ by periodicity. Then f is bounded (in fact $|f(x, y)| \leqslant \frac{1}{2}$) and its discontinuities are at $(0, 0)$ and along the boundaries $x = \pm \pi$, $y = \pm \pi$ (for instance $f(x, \pi) = -f(x, -\pi)$). Hence f is an FC-function.

(ii) If f, g are FC-functions on $\mathbb{T}$ then as has already been pointed out, $h(x, y) = f(x)g(y)$ is an FC-function on $\mathbb{T}^2$. Evidently this example generalizes to a product of p FC-functions on $\mathbb{T}^p$.

(iii) If $f(x, y)$ is defined on $[0, 2\pi] \times [0, 2\pi]$ by

$$f(x, y) = \begin{cases} 1 & \text{if } x + y > 2\pi, \\ 0 & \text{if } x + y \leqslant 2\pi, \end{cases}$$

then f has discontinuities along the edges of the rectangle and along the line $x + y = \pi$, and is thus an FC-function.

It is an easy exercise to show that the set of FC-functions is closed under the formation of sums, products, and translates (regarding the functions as periodic on $\mathbb{R}^2$, or $\mathbb{R}^p$).

Definition 6.3. Let f, g be FC-functions on $\mathbb{T}^2$. The *inner product* (f, g) is defined by

$$(f, g) = \frac{1}{4\pi^2} \iint_{\mathbb{T}^2} f(x, y)\bar{g}(x, y) \, dx \, dy.$$

We say that f, g are *orthogonal* if $(f, g) = 0$. (The definition extends at once to $\mathbb{T}^p$). The sequence (f_n) is *orthogonal* if $(f_n, f_m) = 0$ when $m \neq n$, and *orthonormal* if in addition $(f_n, f_n) = 1$ for all n.

Lemma 6.4 (i) Let $e_{m,n}$ be defined on $\mathbb{T}^2$ by

$$c_{m,n}(x, y) = \exp\{i(mx + ny)\}, \qquad m, n \in \mathbb{Z}.$$

Then the (double) sequence $(e_{m,n})_{m,n=-\infty}^{\infty}$ is orthonormal on $\mathbb{T}^2$.

(ii) Let $\mathbf{n} = (n_1, n_2, \ldots, n_p)$ be an element of $\mathbb{Z}^p$ and $\mathbf{x} = (x_1, x_2, \ldots, x_p)$ be an element of $\mathbb{T}^p$.
We define $e_{\mathbf{n}}(\mathbf{x}) = \exp(i\mathbf{n} \cdot \mathbf{x}) = \exp(i\sum_{j=1}^n n_j x_j)$.
Then the sequence $(e_{\mathbf{n}})_{\mathbf{n} \in \mathbb{Z}^p}$ is orthonormal on $\mathbb{T}^p$.

Proof (i) This result is immediate since the required integral splits (by Theorem A.33 and Corollary A.34) into

$$\int_0^{2\pi} e^{i(m-m')x} \, dx \int_0^{2\pi} e^{i(n-n')y} \, dy$$

which is zero unless $m = m'$, $n = n'$, while if $m = m'$, $n = n'$, both integrals are equal to 2π. The proof of (ii) is similar.

Given this orthonormal system of functions, we can define, as we did in Chapter 1, an expansion of a general function f.

Definition 6.5 (i) The (complex)*Fourier coefficients* of a function f on $\mathbb{T}^2$ are defined for $m, n \in \mathbb{Z}$ by $c_{m,n}(f) = (f, e_{m,n})$

$$= \frac{1}{4\pi^2} \int\int_{\mathbb{T}^2} f(x, y) \exp\{-i(mx + ny)\} \, dx \, dy$$

(note the minus sign in the exponent, resulting from the complex conjugate in Definition 6.3).

(ii) The *Fourier series* of f is the series

$$S(f) = \sum_{m=-\infty}^{\infty} \sum_{n=-\infty}^{\infty} c_{m,n}(f) \exp\{i(mx + ny)\},$$

(at this stage there is naturally no guarantee of any particular type of convergence).

(iii) For an FC-function on $\mathbb{T}^p$, the Fourier coefficients are defined by

$$c_{\mathbf{n}}(f) = (2\pi)^{-p} \int_{\mathbb{T}^p} f(\mathbf{x}) \exp(-i\mathbf{n} \cdot \mathbf{x}) \, d\mathbf{x}, \text{ for } \mathbf{n} \in \mathbb{Z}^p,$$

and the Fourier series is $S(f) = \sum_{\mathbf{n} \in \mathbb{Z}^p} c_{\mathbf{n}}(f) \exp(i\mathbf{n} \cdot \mathbf{x})$.

We can also define real Fourier coefficients with respect to the orthogonal (but not orthonormal) system $\{\cos mx \cos ny, \cos mx \sin ny, \sin mx \cos ny, \sin mx \sin ny\}$, $m, n \geq 0$, but this is more complicated, and we shall not do it except when symmetry properties simplify the situation, as occurs in exercises 1 and 2 at the end of the chapter, for example.

Example 6.6 (i) Let $f(x, y) = g(x)h(y)$, where g, h are FC-functions on $[0, 2\pi]$. Then f is FC and the Fourier coefficients of f are simply the products of the coefficients of g and h:

$$c_{m,n}(f) = \frac{1}{4\pi^2} \int_0^{2\pi} \int_0^{2\pi} g(x)h(y)e^{-imx}e^{-iny} \, dx \, dy$$
$$= c_m(g)c_n(h).$$

It follows that the Fourier series for f is formed by multiplying the series for f and g:

$$S(f) = \sum_{-\infty}^{\infty} c_m(g)e^{imx} \sum_{-\infty}^{\infty} c_n(h)e^{iny},$$

and convergence of $S(f)$ is equivalent to the convergence of both $S(g)$ and $S(h)$.

(This apparently trivial situation covers some useful applications, as the next example shows.)

(ii) Let

$$g(x) = h(x) = \begin{cases} 1, & 0 < x < \pi \\ -1, & -\pi < x < 0 \\ 0 & \text{at } 0, \pm \pi \end{cases}$$

Then we know from Chapter 1 that

$$g(x) = \frac{4}{\pi} \sum_{k=0}^{\infty} \frac{\sin(2k+1)\pi x}{2k+1},$$

where the series is pointwise convergent everywhere, and uniformly convergent on intervals of the form $(\delta, \pi - \delta)$ or $(-\pi + \delta; -\delta)$, $\delta > 0$. It follows that if

$$f(x, y) = \operatorname{sgn}(xy) = \begin{cases} 1, & xy > 0 \\ -1, & xy < 0 \\ 0, & xy = 0 \end{cases} \text{ on } (-\pi, \pi) \times (-\pi, \pi)$$

and $f(x, y) = 0$ if x or $y = \pi$, then $f(x, y) = g(x)g(y)$, and consequently

$$S(f) = \frac{16}{\pi^2} \sum_{k=0}^{\infty} \sum_{j=0}^{\infty} \frac{\sin(2k+1)\pi x}{2k+1} \frac{\sin(2j+1)\pi y}{2j+1}.$$

Again the series is pointwise convergent everywhere, and uniformly convergent on the appropriate rectangles. This example describes how the stress on a uniformly loaded plate supported at its edges is decomposed into its harmonic components. Szilard (1973, sections 1.5 and 1.6) contains several similar examples with their engineering applications. Exercise 10 at the end of the chapter shows how double series can be used to determine the flow of heat in a plate.

(iii) Let $f(x, y) = \min(|x|, |y|)$ for $|x|, |y| \leqslant \pi$.

This is an even function of x and y which is continuous on $\mathbb{T}^2$ but which is not the product of independent functions of x and y as in the examples above. Exercise 1 at the end of the chapter applies and we can write

$$c_{m,n}(f) = \frac{1}{\pi^2} \int_0^\pi \int_0^\pi \min(|x|, |y|) \cos mx \cos ny \, dx \, dy.$$

This interval is evaluated in exercise 3 at the end of the chapter, giving for f the Fourier series

$$S(f) = \frac{\pi}{3} - \frac{2}{\pi} \sum_{m=1}^{\infty} \frac{\cos mx}{m^2} - \frac{2}{\pi} \sum_{n=1}^{\infty} \frac{\cos ny}{n^2} + \frac{2}{\pi} \sum_{n=1}^{\infty} \frac{\cos nx \cos ny}{n^2}.$$

We shall see shortly (Corollary 6.10) that the absolute convergence of this series, together with the continuity of f, is sufficient to ensure that the sum of the series is $f(x, y)$. Exercise 4 at the end of the chapter verifies this result in another way.

144

6.2 CONVERGENCE OF MULTIPLE FOURIER SERIES

We begin our discussion of the convergence of multiple Fourier series by defining the notion of completeness for an orthogonal system of functions, by analogy with Section 1.3.

Definition 6.7 Let f be an FC-function on $\mathbb{T}^p$. We say that f is *almost zero* if $f(x, y) = 0$ at every point of continuity of f.

Evidently this implies that the integral of f over every subrectangle of $\mathbb{T}^p$ is zero.

Definition 6.8 Let (ϕ_n) be an orthogonal sequence of functions on $\mathbb{T}^p$. We say that ϕ_n is *complete* on $\mathbb{T}^p$ if for all FC-functions f, $(f, \phi_n) = 0$ for all n implies that f is almost zero.

We shall be concerned only with the exponential and trigonometric systems on $\mathbb{T}^p$. We show that the exponential system is complete on $\mathbb{T}^2$: the method plainly extends to higher dimensions. The completeness of the associated trigonometric systems is exercise 5 at the end of the chapter.

Theorem 6.9 The exponential system

$$\phi_{m,n}(x, y) = \exp(i(mx + ny)), m, n \in \mathbb{Z}, \text{ is complete on } \mathbb{T}^2.$$

Proof Let f be a FC-function for which $(f, \phi_{m,n}) = 0$ for all m, n.

From the definition of an FC-function, there are a fixed finite number of lines on which f may have discontinuities: let $x = x_1, x_2, \ldots, x_k$ be the lines parallel to the y-axis (other discontinuities will have no effect on the argument).

Define $g_n(x) = \int_0^{2\pi} f(x, y)e^{-iny} \, dy$: g_n is an FC-function of x with possible discontinuities only at points of $X = \{x_1, x_2, \ldots, x_k\}$.

The hypothesis says that $\int_0^{2\pi} g_n(x)e^{-imx} \, dx = 0$ for all m, and hence that for each n, $g_n(x)$ is almost zero by Theorem 1.10(i). In particular, we know that $g_n(x) = 0$ for all n and all x not in X. Hence by Theorem 1.10(i) again, for all x not in X, $f(x, y)$ is almost zero as a function of y. In particular, f must be zero at all points (x, y) with x not in X at which f is continuous, which is sufficient to show that f is almost zero on $\mathbb{T}^2$, as required.

Corollary 6.10 Let f be continuous on $\mathbb{T}^2$ and let $S(f)$ be uniformly convergent on $\mathbb{T}^2$. Then $S(f)$ converges to f on $\mathbb{T}^2$.

Proof Let g be the sum of $S(f)$: g is continuous by uniform convergence. Also by uniform convergence, we can integrate termwise to show that $c_{m,n}(f) = c_{m,n}(g)$ and hence that $c_{m,n}(f - g) = 0$. Since $f - g$ is continuous, it must vanish identically by Theorem 6.9, and the result follows.

Note Evidently Theorem 6.9 and Corollary 6.10 extend immediately to any number of dimensions. Results such as Bessel's equation can be proved in exactly the same way as in Chapter 1: this is done in exercise 6.

We now consider the somewhat harder problem of finding a sufficient condition on f alone which will guarantee convergence of $S(f)$. We begin by showing, as was mentioned in the introductory paragraph, that no purely local condition can be sufficient.

Theorem 6.11 There exists an FC-function which is zero in a neighbourhood of the origin in $\mathbb{T}^2$, whose Fourier series diverges there.

Proof We recall the construction (Theorem 2.16) of a continuous function whose Fourier series is divergent. In that theorem we defined

$$S_n(x) = \sum_{j=1}^{n} \frac{\sin jx}{j}, \quad \text{and} \quad F(N, n, x) = 2(\sin Nx)S_n(x).$$

We then chose rapidly increasing sequences $n_r = 2^{r^2}$, $N_r = 2n_r$, and showed that

$$f(x) = \sum_{r=1}^{\infty} \frac{1}{r^2} F(N_r, n_r, x)$$

has the required property.

If we choose still more rapidly increasing sequences, for instance $N_r = 2n_r = 2 \cdot 2^{r^3}$, then

$$\frac{1}{r^2} \sum_{j=1}^{n_r} \frac{1}{j} > \frac{1}{r^2} \log(n_r) > r \log 2$$

is unbounded, and the argument in Theorem 2.16 shows that the partial sums of the Fourier series of f are unbounded at $x = 0$.

Now define $g(y) = 0$ on $(-\pi/2, \pi/2)$, and 1 elsewhere, so that

$$S(g)(y) = \frac{1}{2} - \frac{2}{\pi} \sum_{k=0}^{\infty} (-1)^k \frac{\cos(2k+1)\pi y}{2k+1}.$$

Let $h(x, y) = f(x)g(y)$: clearly h is zero in a neighbourhood of $(0, 0)$, and

$$S_{m_1, n_1}(h)(x, y) = \sum_{-m_1}^{m_1} c_m(f)e^{imx} \sum_{-n_1}^{n_1} c_n(g)e^{iny} = S_{m_1}(f)S_{n_1}(g).$$

Since $S_{n_1}(g)$ is never zero at $y = 0$ (π is irrational), this expression is unbounded for fixed n_1, as $m_1 \to \infty$. Hence $S_{m_1, n_1}(h)$ does not have a limit as $m_1, n_1 \to \infty$; in other words, the Fourier series of h is divergent at $(0, 0)$.

For our main theorem on convergence, we need the notion of total variation for a piecewise monotone function.

Definition 6.12

(i) Let f be a bounded monotone function on (a, b), so that the limits $l_1 = \lim_{x \to a_+} f(x)$ and $l_2 = \lim_{x \to b_-} f(x)$ exist.

We define the *total variation* of f on (a, b) as $|l_1 - l_2|$.

(ii) Let f be a bounded piecewise monotone function on (a, b). Then the *total variation* of f on (a, b), which we denote by $V_f(a, b)$ is the sum of the total variations over the subintervals on which f is monotone.

(This is not quite the standard definition of total variation – but it will suffice for our purposes.)

The next lemma shows how, if f is piecewise monotone, we can get a bound on integrals of the form $\int_{-\pi}^{\pi} f(x)g(x)\,dx$.

Lemma 6.13 Let f be bounded and piecewise monotone on $[-\pi, \pi]$ and g be a FC-function. Let

$$M_1 = \sup \{|f(x)|; |x| \leqslant \pi\}, \text{ and}$$

$$M_2 = \sup \left\{ \left| \int_0^x g(t)dt \right|; |x| \leqslant \pi \right\}.$$

Then
$$\left| \int_{-\pi}^{\pi} f(x)g(x)dx \right| \leqslant 2M_2(V_f(-\pi, \pi) + M_1).$$

Proof Let $G(x) = \int_0^x g(t)dt$, so that $|G(x)| \leqslant M_2$. Suppose that f is positive on $[-\pi, \pi]$ and increasing on (a, b). Then by Lemma 2.10(ii),

$$\left| \int_a^b f(x)g(x)dx \right| = \left| f(b-) \int_c^b g(x)dx \right| \leqslant V_f(a, b)2M_2.$$

Hence by addition over all subintervals on which f is monotone,

$$\left| \int_{-\pi}^{\pi} f(x)g(x)\,dx \right| \leqslant 2M_2 V_f(-\pi, \pi) \text{ when } f \text{ is positive.}$$

In the general case when f is not necessarily positive, we let $f_1 = f + M_1$ which is positive, and for which $V_{f_1} = V_f$.

Then the above argument shows that

$$\left| \int_{-\pi}^{\pi} f(x)g(x)dx \right| \leqslant \left| \int_{-\pi}^{\pi} f_1(x)g(x)\,dx \right| + M_1 \left| \int_{-\pi}^{\pi} g(x)dx \right|$$
$$\leqslant 2M_2 V_{f_1}(-\pi, \pi) + 2M_1 M_2 = 2M_2(V_f(-\pi, \pi) + M_1), \text{ as required.}$$

Lemma 6.13 gives us the crucial step in our pointwise convergence theorem. For simplicity we restrict ourselves to two variables.

Theorem 6.14 Let f be an FC-function on $\mathbb{T}^2$ which has the following properties:
(i) there is a constant M for which
   (a) $|f(x, b) - f(a, b)| \leqslant M|x - a|$, and

(b) $|f(x, y) - f(x, b)| \leq M|y - b|$ for all x and y.
(ii) For each fixed y, the function $\phi_y(x) = f(x, y) - f(x, b)$ is piecewise monotone, and there is a constant M_2 for which $V_{\phi_y}(-\pi, \pi) \leq M_2|y - b|$ for all y.
Then the Fourier series of f is convergent at (a, b), with sum $f(a, b)$.

(Note that (i) provides Lipschitz continuity of f in the neighbourhood of the line $y = b$, while (ii) does the same for the total variation of ϕ_y.)

Proof We take $(a, b) = (0, 0)$ without loss of generality. Denote by $S_{m,n}f(x, y)$ the partial sum

$$\sum_{r=-m}^{m} \sum_{s=-n}^{n} c_{r,s}(f)e^{i(rx+sy)}$$

of the Fourier series of f.
It follows as in Lemma 2.1 that

$$S_{m,n}f(x, y) = \frac{1}{4\pi^2} \int_{-\pi}^{\pi} \int_{-\pi}^{\pi} f(u, v)D_m(x - u)D_n(y - v)\, du\, dv,$$

where D_m is the Dirichlet kernel.

In particular, $S_{m,n}f(0,0) - f(0,0) = \frac{1}{4\pi^2}\int_{-\pi}^{\pi}\int_{-\pi}^{\pi}(f(u,v) - f(0,0))D_m(u)D_n(v)\,du\,dv$,

and we have to show that this integral tends to zero as $m, n \to \infty$.
We write this integral in the form (replacing (u, v) by (x, y))

$$\frac{1}{4\pi^2} \int_{-\pi}^{\pi} \int_{-\pi}^{\pi} (f(x, y) - f(x, 0))D_m(x)D_n(y)\, dx\, dy$$

$$+ \frac{1}{4\pi^2} \int_{-\pi}^{\pi} \int_{-\pi}^{\pi} (f(x, 0) - f(0, 0))D_m(x)D_n(y)\, dx\, dy$$

$$= I_1 + I_2, \text{ say.}$$

The integral I_2 is easily dealt with: since

$$\frac{1}{2\pi} \int_{-\pi}^{\pi} D_n(y)\, dy = 1,$$

it is simply

$$\frac{1}{2\pi} \int_{-\pi}^{\pi} (f(x, 0) - f(0, 0))D_m(x)\,dx = \frac{1}{2\pi} \int_{-\pi}^{\pi} \frac{f(x, 0) - f(0, 0)}{\sin \frac{1}{2}x} \sin(n + \tfrac{1}{2})x\, dx$$

which tends to zero by the Riemann–Lebesgue lemma using the hypothesis (i)(a), as in the proof of Theorem 2.3(i).
We can use the same argument to show that I_1 tends to zero if we can show that the function of y given by

$$\frac{1}{2\pi} \int_{-\pi}^{\pi} (f(x, y) - f(x, 0))D_m(x)\, dx$$

is Lipschitz continuous at $y = 0$: that is, we require the existence of a constant C for which

$$\left| \frac{1}{2\pi} \int_{-\pi}^{\pi} \phi_y(x) D_m(x)\, dx \right| \leqslant C|y|, \text{ where } \phi_y(x) = f(x, y) - f(x, 0).$$

We showed in Lemma 2.11 that the integrals $\int_0^x D_m(t)\, dt$ are uniformly bounded: there is a constant M_2 for which $|\int_0^x D_m(t)\, dt| \leqslant M_2$ for all x and m. This enables us to apply Lemma 6.13 with $f(x) = \phi_y(x)$ and $g(x) = D_m(x)$, to obtain

$$\left| \int_{-\pi}^{\pi} \phi_y(x) D_m(x)\, dx \right| \leqslant 2M_2(V_{\phi_y}(-\pi, \pi) + M_1)$$

where

$$M_1 = \sup \{\phi_y(x);\ -\pi \leqslant x \leqslant \pi\}.$$

The hypotheses (i)(b) and (ii) now show that M_1 and V_{ϕ_y} are both bounded by a constant multiple of $|y|$, and this completes the proof.

The use of the conditions 6.14(i) and (ii) is best illustrated by means of examples.

Example 6.15 (i) Let $f(x, y) = \min(|x|, |y|)$ for $|x|, |y| \leqslant \pi$. (This is the example considered by other means in Example 6.6(iii).) We will show convergence at $(0, 0)$.

Evidently the conditions (i)(a) and (b) are satisfied with $M = 1$. For $y \neq 0$, the function $\phi_y(x)$ is simply $f(x, y)$, and this is equal to $|y|$ if $|x| \geqslant |y|$, or to $|x|$ if $|x| < |y|$.

Hence $V_{\phi_y}(-\pi, \pi) = 2|y|$, and hypothesis (ii) is satisfied.

It is easily seen that the hypotheses of Theorem 6.14 are in fact satisfied for the above function in relation to all points of $\mathbb{T}^2$ and that consequently the Fourier series is convergent to f everywhere, as we found in Example 6.6(iii).

The same hypotheses as in Theorem 6.14, if valid uniformly over a rectangle in $\mathbb{T}^2$, supply a test for uniform convergence over the rectangle, as may be proved by methods analogous to those used to prove uniform convergence in Chapter 2, but we shall not carry this out here.

(ii) Let

$$f(x, y) = \begin{cases} xy & \text{for } |x| + |y| \leqslant \pi, \\ 0 & \text{elsewhere with } |x|, |y| \leqslant \pi \end{cases}$$

This is evidently an FC-function, having discontinuities only along the lines $\pm x \pm y = \pi$.

We shall show that its Fourier series converges at $(0, 0)$. Note that the discontinuities prevent uniform convergence of the series, and so we cannot appeal to Theorem 6.10. For this function we have $f(0, y) = f(x, 0) = 0$, and $|f(x, y)| \leqslant \pi|y|$, so that the condition 6.14(i) is immediate.

The total variation of $\phi_y(x) = f(x, y)$ on $(-\pi, \pi)$ is $2V_{\phi_y}(-|y|, |y|) = 4y^2$ so that 6.14 (ii) is also satisfied, and the series is convergent at $(0, 0)$.

We shall not attempt to investigate convergence at other points.

6.3 MULTIPLE FOURIER INTEGRALS

When we come to consider the Fourier analysis of functions which are defined (non-periodically) on the whole of $\mathbb{R}^p$, we extend our notion of an FV-function from one to several variables using the notion of an FC-function on a bounded set. To be precise, we adopt the following definition, which is analogous to Definition A.13.

Definition 6.16 Let f be a real- or complex-valued function on $\mathbb{R}^p$ and suppose that

(a) f is an FC-function on every bounded rectangle R in $\mathbb{R}^p$, and
(b) for some constant M, $\int_R |f(x_1,\ldots,x_p)| \, dx_1,\ldots dx_p \leqslant M$ for every such R.
Then we say that f is an *FV-function* on $\mathbb{R}^p$.

It is then easy to show that the integral of f over $\mathbb{R}^p$ is determined as the limit of integrals over bounded rectangles, as their sides tend (separately) to infinity: We write $\int_{\mathbb{R}^p} f(x_1,\ldots,x_p) \, dx_1 \ldots dx_p$ or simply $\int_{\mathbb{R}^p} f$ for its value.

Example 6.17 (i) Let f be any function which is continuous on a closed, bounded rectangle $R \subseteq \mathbb{R}^p$ and zero elsewhere. Then f is evidently an FV-function and $\int_R f = \int_{\mathbb{R}^p} f$.

(ii) Let f be any function which depends only on $|x| = r$, the distance of the point x from the origin. Such a function is called *radial*: we shall see several examples of them shortly. We can put $f(x) = \phi(r)$ and a change to polar co-ordinates shows that f is an FV-function if and only if ϕ is bounded and $\phi(r)r^{p-1}$ is an FV-function on $(0, \infty)$: in fact $\int_{\mathbb{R}^p} f = \omega_{p-1} \int_0^\infty \phi(r)r^{p-1} \, dr$, where ω_{p-1} is defined in exercise 9 at the end of this chapter.

For instance consider the following examples:

(a) Let $f(x) = (1 + x_1^2 + x_2^2 + \cdots + x_p^2)^{-\alpha}$, $\alpha > 0$.

Evidently f is radial with $\phi(r) = (1 + r^2)^{-\alpha}$, so f is an FV-function provided

$$\int_0^\infty \frac{r^{p-1}}{(1+r^2)^\alpha} \, dr$$

converges, i.e. if $2\alpha > p$. In particular, if $p = 2$, $f(x, y) = (1 + x^2 + y^2)^{-\alpha}$ is an FV-function for $\alpha > 1$.

(b) Let $f(x) = e^{-a|x|}$ or $e^{-a|x|^2}$ for $a > 0$.

Evidently $\int_0^\infty r^{p-1} e^{-ar} \, dr$ and $\int_0^\infty r^{p-1} e^{-ar^2} \, dr$ are finite, and so both $e^{-a|x|}$ and $e^{-a|x|^2}$ are FV-functions.

We shall shortly find the Fourier transform of all these functions. We define the Fourier transform on $\mathbb{R}^p$ by analogy with the Definition 5.6 on $\mathbb{R}$.

Definition 6.18 Let f be an FV-function on $\mathbb{R}^p$. Then for $t \in \mathbb{R}^p$ we define the Fourier transform

$$\hat{f}(\mathbf{t}) = \int_{\mathbb{R}^p} f(\mathbf{x}) \exp(-2\pi i \mathbf{x} \cdot \mathbf{t}) \, d\mathbf{x}$$

where $\mathbf{x}\cdot\mathbf{t}$ denotes the scalar product $\sum_{k=1}^{p} x_k t_k$ of $\mathbf{x}$ and $\mathbf{t}$. In particular, when $p = 2$, and we write (x, y) for $\mathbf{x}$ and (t, u) for $\mathbf{t}$, we have $\hat{f}(t, u) = \int_{\mathbb{R}^2} f(x, y) e^{-2\pi i (xt + yu)} \, dx \, dy$.

Since the exponential factor has modulus one, the integral exists for all $\mathbf{y}$ in $\mathbb{R}^p$. The formal properties of f are identical with those listed in Lemma 5.7 and we shall not restate them: as an example, notice that if $\mathbf{h} \in \mathbb{R}^p$ and $f_{\mathbf{h}}(\mathbf{x}) = f(\mathbf{x} - \mathbf{h})$ is the translate of f by $\mathbf{h}$, then $\hat{f}_{\mathbf{h}}(\mathbf{y}) = \exp(-2\pi i \mathbf{h}\cdot\mathbf{y})\hat{f}(\mathbf{y})$. Similarly, the uniform continuity of $\hat{f}$ and the fact that $\hat{f}(\mathbf{y}) \to 0$ as $|\mathbf{y}| \to \infty$ is proved as in Theorem 5.8(iii).

Example 6.19 For our first example of a Fourier integral we take the Gauss function $f(\mathbf{x}) = e^{-a|\mathbf{x}|^2} = \exp(-a\sum_{k=1}^{p} x_k^2)$.

We have $\hat{f}(\mathbf{t}) = \int_{\mathbb{R}^p} \exp(-a\sum_{k=1}^{p} x_k^2 - 2\pi i \sum_{k=1}^{p} x_k t_k) \, dx_1 \dots dx_p$, and evidently the integral splits into the products of p integrals of the form $\int_{-\infty}^{\infty} \exp(-a x_k^2 - 2\pi i x_k t_k) \, dx_k$, which is equal to

$$\sqrt{\frac{\pi}{a}} \exp(-\pi^2 t_k^2 / a)$$

as shown in Example 5.9(i).

Consequently

$$\hat{f}(\mathbf{t}) = (\pi/a)^{p/2} \exp\left(-\frac{\pi^2}{a}|t|^2\right):$$

in particular when $a = \pi$, f and $\hat{f}$ have the same functional form.

This example also shows that if we define the inversion integral as we did in Chapter 5, by $\int_{-\infty}^{\infty} \hat{f}(t) e^{2\pi \cdot x \cdot t} \, dt$, then for this function at least we have

$$f(\mathbf{x}) = \int_{-\infty}^{\infty} \hat{f}(t) e^{2\pi i x \cdot t} \, dt$$

and the inversion formula is verified in this case.

Example 6.20 Let $f(\mathbf{x}) = e^{-a|\mathbf{x}|}$, $a > 0$: we shall show that in two dimensions we have

$$\hat{f}(\mathbf{t}) = \frac{2\pi a}{(a^2 + |\mathbf{t}|^2)^{3/2}}.$$

We have $\hat{f}(u, v) = \int_{-\infty}^{\infty}\int_{-\infty}^{\infty} e^{-a(x^2 + y^2)^{1/2} - 2\pi i (xu + yv)} \, dx \, dy$ and on changing to polar co-ordinates $x + iy = re^{i\theta}$, $u + iv = Re^{i\phi}$, the integral becomes

$$\int_0^{\infty}\int_0^{2\pi} e^{-ar - 2\pi i R r \cos(\theta - \phi)} \, d\theta r dr = \int_0^{2\pi}\left\{\int_0^{\infty} e^{-(a + 2\pi i R \cos\theta)r} r \, dr\right\} d\theta$$

since the integral is independent of the value of ϕ. We can evaluate the inner integral since $\int_0^\infty x e^{-cx} dx = 1/c^2$ for $\text{Re}\, c > 0$, and we obtain

$$\hat{f}(Re^{i\phi}) = \int_0^{2\pi} \frac{d\theta}{(a + 2\pi i R \cos\theta)^2}.$$

To evaluate this integral, observe that for $a > |b|$, we have

$$\int_0^\pi \frac{d\theta}{a + b\cos\theta} = \int_0^\infty \frac{2dt}{(a+b) + (a-b)t^2} \qquad (\text{putting } t = \tan\tfrac{1}{2}\theta)$$

$$= \frac{\pi}{\sqrt{(a^2 - b^2)}}$$

and thus on differentiating with respect to a,

$$\int_0^\pi \frac{d\theta}{(a + b\cos\theta)^2} = \frac{\pi a}{(a^2 - b^2)^{3/2}}.$$

Hence

$$\hat{f}(Re^{i\phi}) = \frac{2\pi a}{(a^2 + 4\pi^2 R^2)^{3/2}},$$

or

$$\hat{f}(\mathbf{t}) = \frac{2\pi a}{(a^2 + 4\pi^2 |\mathbf{t}|^2)}, \text{ as required.}$$

We note (without proof) the corresponding formula in $\mathbb{R}^p$: if $f(\mathbf{x}) = e^{-2\pi a|\mathbf{x}|}$, then

$$\hat{f}(\mathbf{t}) = \frac{c_p a}{(a^2 + |\mathbf{t}|^2)^{\frac{1}{2}(p+1)}}$$

where $2/c_p = \omega_p$, the surface of the unit sphere in $\mathbb{R}^{p+1}$ (see exercise 8). Details of this calculation can be found in Stein and Weiss (1971).

We now consider circumstances in which the Fourier integral can be inverted by the formula $f(\mathbf{x}) = \int_{\mathbb{R}^p} \hat{f}(\mathbf{t}) \exp(2\pi i \mathbf{t} \cdot \mathbf{x}) d\mathbf{t}$.

The method is parallel to that used in Section 5.3, and we shall outline it informally here, rather than stating the results as a sequence of theorems.

We notice first that Example 6.19 shows that the Gauss function $\phi_t(\mathbf{x}) = \exp(-t\sum_{j=1}^p x_j^2)$, $t > 0$, is available to use in the definition of summability of integrals: we shall say that the integral $\int_{\mathbb{R}^p} f(\mathbf{x}) d\mathbf{x}$ is *G-summable* with value A if $\int_{\mathbb{R}^p} \phi_t(\mathbf{x}) f(\mathbf{x}) d\mathbf{x} \to A$ as $t \to 0_+$.

Also, the result of Theorem 5.13 remains true (with the same proof): for an FV-function f,

$$f * \hat{\phi}_t(\mathbf{x}) = \int_{\mathbb{R}^p} \phi_t(\mathbf{y}) \hat{f}(\mathbf{y}) \exp(2\pi i \mathbf{x} \cdot \mathbf{y}) d\mathbf{y},$$

where we restrict use of the formula to the case when ϕ_t is the Gauss function, so that, in particular, ϕ_t is even.

The limiting value of the right-hand side, as $t \to 0_+$, is by definition the G-value of the integral $\int_{\mathbb{R}^p} \hat{f}(\mathbf{y}) \exp(2\pi i \mathbf{x} \cdot \mathbf{y}) d\mathbf{y}$: one shows as in Theorem 5.16 that the G-value is the same as the ordinary value, when $\hat{f}$ is an FC-function.

Consequently it remains to show that, as $t \to 0_+$, $f * \hat{\phi}_t(\mathbf{x}) \to f(\mathbf{x})$ in some suitable sense – for instance, at points of continuity of an FV-function. As in Theorem 5.17, we need only the properties (a), (b), (c) listed there, namely (for $g_t = \hat{\phi}_t$):

(a) $\hat{\phi}_t(\mathbf{x}) \geqslant 0$ for all $\mathbf{x}$;
(b) $\int_{\mathbb{R}^p} \hat{\phi}_t(\mathbf{x}) d\mathbf{x} = 1$;
(c) for any $\delta > 0$, $\int_{|\mathbf{x}| \geqslant \delta} \hat{\phi}_t(\mathbf{x}) d\mathbf{x} \to 0$ as $t \to 0_+$.

From Example 6.19 we know that

$$\hat{\phi}_t(\mathbf{y}) = \left(\frac{\pi}{t}\right)^{p/2} \exp\left(-\frac{\pi^2}{t}|\mathbf{y}|^2\right), \quad t > 0, \ \mathbf{x}, \mathbf{y} \in \mathbb{R}^p,$$

from which (a) is immediate, (b) follows since the inversion formula is valid for ϕ_t, and (c) follows since the integral is

$$c \int_\delta^\infty \hat{\phi}_t(r) r^{p-1} dr \qquad (r = |\mathbf{x}|)$$

$$= c \int_\delta^\infty \left(\frac{\pi}{t}\right)^{p/2} \exp\left(-\frac{\pi^2}{t}r^2\right) r^{p-1} dr$$

$$= c \left(\frac{\pi}{t}\right)^{p/2} \int_{\pi^2\delta^2/t}^\infty e^{-u} \left(\frac{tu}{\pi^2}\right)^{p/2} \frac{du}{2u}, \qquad \text{putting } \frac{\pi^2}{t}r^2 = u,$$

$$= \frac{c}{2\pi^{p/2}} \int_{\pi^2\delta^2/t}^\infty e^{-u} u^{(1/2)p-1} du, \qquad \text{which tends to zero as } t \to 0_+$$

We have thus arrived at the following result.

Theorem 6.21 Let f be a bounded FV-function on $\mathbb{R}^p$.

Then $\int_{\mathbb{R}^p} \hat{f}(\mathbf{y}) \exp(-t|\mathbf{y}|^2 + 2\pi i \mathbf{x} \cdot \mathbf{y}) d\mathbf{y} \to f(\mathbf{x})$ as $t \to 0_+$, at each point of continuity of f.

In particular, we have the corollaries

(i) if f and $\hat{f}$ are bounded continuous FV-functions, then $f(\mathbf{x}) = \int_{\mathbb{R}^p} \hat{f}(\mathbf{y}) \exp(2\pi i \mathbf{x} \cdot \mathbf{y}) d\mathbf{y}$ for all $\mathbf{x}$, and

(ii) if f is a bounded FV-function which is continuous at 0, and $\hat{f} \geqslant 0$, then $\hat{f}$ is an FV-function, and

$$f(0) = \int_{\mathbb{R}^p} \hat{f}(\mathbf{y}) d\mathbf{y}.$$

The second corollary may be made the basis for the extension of the Fourier transform to square-integrable functions as in Section 5.4.

Instead of repeating these formal deductions, we conclude this chapter with an example on the conduction of heat, which recalls the starting point of our theory, in Chapter 1.

Example 6.22 Consider an infinite sheet of a uniform conducting material, with initial temperature given by $f_0(\mathbf{x})$, $\mathbf{x} \in \mathbb{R}^2$. We shall find the temperature in the sheet at subsequent times t. (Notice that a similar calculation will go through in dimensions greater than 2, with a suitably amended physical interpretation.)

We let $f(\mathbf{x}, t)$ be the temperature at $\mathbf{x} = (x_1, x_2) \in \mathbb{R}^2$ and time $t \geqslant 0$: the equation governing the flow of heat is

$$\frac{\partial^2 f}{\partial x_1^2} + \frac{\partial^2 f}{\partial x_2^2} = \frac{1}{k^2} \frac{\partial f}{\partial t},$$

for constant k.

As usual we seek to separate the variables by putting $f(x_1, x_2, t) = g(x_1, x_2)h(t)$, from which we obtain

$$\frac{1}{g}\left(\frac{\partial^2 g}{\partial x_1^2} + \frac{\partial^2 g}{\partial x_2^2}\right) = \frac{1}{k^2} \frac{h'(t)}{h(t)},$$

so that $h(t) = e^{-ct}$ for some positive c.

We then have

$$\frac{\partial^2 g}{\partial x_1^2} + \frac{\partial^2 g}{\partial x_2^2} + \frac{c}{k^2} g = 0,$$

which is satisfied by functions of the form

$$g(x_1, x_2) = e^{i(ax_1 + bx_2)} \text{ provided that } a^2 + b^2 = c/k^2.$$

This gives solutions of the form

$$f(\mathbf{x}, t) = \exp\{i(ax_1 + bx_2) - k^2(a^2 + b^2)t\},$$

and if we allow a, b to assume all real values, then we obtain for $f(\mathbf{x}, t)$ an integral of the form

$$f(\mathbf{x}, t) = \int_{\mathbb{R}^2} F(y_1, y_2) \exp\{2\pi i(x_1 y_1 + x_2 y_2) - 4\pi^2 k^2(y_1^2 + y_2^2)t\} \, d\mathbf{y}$$

where the function F gives the weight corresponding to each pair $(y_1, y_2) = \mathbf{y}$.

In particular, putting $t = 0$, we see that

$$f_0(\mathbf{x}) = \int_{\mathbb{R}^2} F(\mathbf{y}) \exp\{2\pi i \mathbf{x} \cdot \mathbf{y}\} \, d\mathbf{y}$$

so that f_0 must be the inverse Fourier transform of F. Hence if f_0 satisfies the conditions of the inversion theorem 6.21, we have $F(\mathbf{y}) = \hat{f}_0(\mathbf{y})$.

We can also see that if

$$G_t(x_1, x_2) = \frac{1}{4\pi k^2 t} \exp\left\{-\frac{1}{4k^2 t}(x_1^2 + x_2^2)\right\}$$

then $\hat{G}_t(y_1, y_2) = \exp\{-4\pi^2 k^2 t(y_1^2 + y_2^2)\}$ (Example 6.19).

Thus we can write

$$f(\mathbf{x}, t) = \int_{\mathbb{R}^2} \hat{f}_0(\mathbf{y}) \hat{G}_t(\mathbf{y}) \exp\{2\pi i (x_1 y_1 + x_2 y_2)\} \, d\mathbf{y}.$$

Since the Fourier transform of the convolution $f_0 * G_t$ is the product $\hat{f}_0 \hat{G}_t$, it follows that we have simply

$$f(\mathbf{x}, t) = (f_0 * G_t)(\mathbf{x}),$$

which gives the general form of the solution for arbitrary f_0.

Finally, we specialize further to the case when f_0 too is given by an exponential function $f_0(x) = e^{-a|x|^2}$, $a > 0$.

We can evaluate the convolution explicitly, since if $f_1(t) = e^{-at^2}$, $f_2(t) = e^{-bt^2}$, then

$$f_1 * f_2(t) = \sqrt{\frac{\pi}{a+b}} \exp\left\{ -\frac{abt^2}{a+b} \right\}.$$

We have

$$f(\mathbf{x}, t) = \frac{1}{4\pi k^2 t} \frac{\pi}{(a + 1/4k^2 t)} \exp\left\{ -\frac{(a/4k^2 t)}{(a + 1/4k^2 t)} (x_1^2 + x_2^2) \right\}$$

$$= \frac{1}{1 + 4ak^2 t} \exp\left\{ \frac{-a}{1 + 4ak^2 t} |\mathbf{x}|^2 \right\}.$$

Evidently this gives $f_0(\mathbf{x}) = e^{-a|\mathbf{x}|^2}$ when $t = 0$, and, not surprisingly, it also tends uniformly to zero as $t \to \infty$.

It is more interesting to ask how such an initial distribution of temperature could have arisen. For this we consider negative values of t (we 'run time backwards', so to speak) and observe that as $t \to -1/4ak^2$, the temperature at the origin tends to infinity, while for all other $\mathbf{x}$, $|\mathbf{x}| > 0$, and so the temperature for other $\mathbf{x}$ tends rapidly to zero. It is easily checked that the total amount of heat ($\int_{\mathbb{R}^2} f(\mathbf{x}, t) d\mathbf{x}$) is constant, so we have a situation in which at time $t = -1/4k^2 a$ all the heat was concentrated at the origin (a so-called 'point-source' of heat): this heat was then dispersed through the plane according to the distribution $f(\mathbf{x}, t)$ already found. These point sources bring us to the beginning of the theory of generalized functions, or distributions, and make a suitable point to finish this elementary survey of Fourier theory, and refer the reader to the more advanced literature contained in Appendix C.

EXERCISES

1 Let f be an even function on $\mathbb{T}^2$, i.e. $f(-x, -y) = f(-x, y) = f(x, y)$ for all x, y. Show that $c_{m,n}(f) = c_{-m,n}(f) = c_{-m,-n}(f) = 1/\pi^2 \int_0^\pi \int_0^\pi f(x, y) \cos mx \cos ny \, dx \, dy$, and that $S(f) = c_{00}(f) + 2\sum_{m=1}^\infty c_{m,0}(f) \cos mx + 2\sum_{n=1}^\infty c_{0,n}(f) \cos ny + 4\sum_{n=1}^\infty \sum_{m=1}^\infty c_{m,n}(f) \cos mx \cos ny$.

2 Frame similar results to those of exercise 1 for functions which satisfy

(a) $f(-x, -y) = -f(x, -y) = f(x, y)$,
(b) $f(-x, -y) = f(x, -y) = -f(x, y)$, on $\mathbb{T}^2$.

(*Hint*: functions satisfying (a) have a Fourier series consisting wholly of sine terms, while those satisfying (b) have cosine terms in x, and sine terms in y.)

3 Show that

$$\frac{1}{\pi^2} \int_0^\pi \left(\int_0^x y \cos ny\, dy \right) \cos mx\, dx = \begin{cases} \pi/6, & m = n = 0 \\ (-1)^m/\pi m^2, & n = 0, m \geqslant 1 \\ -\dfrac{(1+(-1)^n)}{\pi n^2}, & m = 0, n \geqslant 1 \\ 1/4n^2\pi, & m = n \geqslant 1 \\ (-1)^{m+n-1}/\pi(n^2-m^2), & m \neq n, m, n \geqslant 1. \end{cases}$$

Deduce that the Fourier coefficients of the function in Example 6.6(iii) are given by

$$c_{m,n}(f) = \begin{cases} \pi/3 & m = n = 0 \\ -1/\pi m^2, & n = 0, m \geqslant 1 \\ -1/\pi n^2, & m = 0, n \geqslant 1 \\ 1/2\pi n^2, & m = n \geqslant 1 \\ 0 & \text{otherwise.} \end{cases}$$

4 Show that for $|x| \leqslant 2\pi$.

$$\sum_{n=1}^\infty \frac{\cos nx}{n^2} = \frac{\pi^2}{6} - \frac{1}{2}\pi|x| + \tfrac{1}{4}x^2.$$

Deduce that in Example 6.6(iii), the sum of the series is $\min(|x|, |y|)$ for $|x|$, $|y| \leqslant \pi$.

(*Hints*: (i) $2 \cos nx \cos ny = \cos n(x+y) + \cos n(x-y)$,
        (ii) for positive a, b, $\min(a, b) = \tfrac{1}{2}(a+b-|a-b|)$.)

5 Deduce from Theorem 6.9 the completeness of the trigonometric system $\{\cos mx \cos ny,\ \cos mx \sin ny,\ \sin mx \cos ny,\ \sin mx \sin ny\}_{m,n \geqslant 0}$ on $\mathbb{T}^2$.

6 Let f be an FC-function on $\mathbb{T}^2$, with Fourier coefficients $c_{m,n}$, $-\infty < m, n < \infty$ and let $S_{m,n}$ be the partial sum of its Fourier series. Show as in Section 1.5 that

$$\frac{1}{4\pi^2} \int_0^{2\pi} \int_0^{2\pi} |f(x, y)|^2\, dx\, dy = \sum_{-\infty}^\infty \sum_{-\infty}^\infty |c_{m,n}|^2$$

and hence that

$$\int_0^{2\pi} \int_0^{2\pi} |S_{m,n}(x, y) - f(x, y)|^2\, dx\, dy \to 0 \text{ as } m, n \to \infty.$$

7 Which hypothesis of Theorem 6.14 is not satisfied in the example of Theorem 6.11?

8 Show that the intersection of the unit sphere in $\mathbb{R}^n$ with an $(n-1)$-dimensional

subspace, at distance r from the centre, is a sphere in $\mathbb{R}^{n-1}$ with radius $(1 - r^2)^{1/2}$. Deduce that if V_n denotes the volume of the unit sphere in $\mathbb{R}^n$, then

$$V_n = V_{n-1} \int_{-1}^{1} (1 - r^2)^{(1/2)(n-1)} \, dr.$$

Use the results of Lemma B.1 to deduce that for $n \geqslant 2$,

$$V_n = \frac{2\pi}{n} V_{n-2}.$$

Use the known values $V_1 = 2$, $V_2 = \pi$ (or $V_0 = 1$) to show that

$$V_n = \begin{cases} \dfrac{\pi^{1/2n}}{(\frac{1}{2}n)!} & \text{if } n \text{ is even,} \\[2ex] \dfrac{2(2\pi)^{(1/2)(n-1)}}{1 \cdot 3 \cdot 5 \cdots n} & \text{if } n \text{ is odd.} \end{cases}$$

9   Let S_{n-1} denote the surface of the unit sphere in $\mathbb{R}^n$. Show that S_{n-1} has $((n-1)$-dimensional) volume $\omega_{n-1} = nV_n$, where V_n is found from exercise 8.

10 (i) Show that the heat equation

$$\frac{\partial^2 f}{\partial x^2} + \frac{\partial^2 f}{\partial y^2} = \frac{1}{k} \frac{\partial f}{\partial t}$$

is satisfied by $f(x, y, t) = g(x, y)h(t)$ where $h(t) = e^{-ct}$, and

$$\frac{\partial^2 g}{\partial x^2} + \frac{\partial^2 g}{\partial y^2} + \frac{c}{k} g = 0.$$

(ii) Consider a uniform rectangular plate of a conducting material with sides of length a, b. Let $f(x, y, t)$ denote the temperature at a point (x, y), $0 \leqslant x \leqslant a$, $0 \leqslant y \leqslant b$, $t \geqslant 0$.

Find $f(x, y, t)$ in the following two cases:

(a) there is an initial temperature distribution $f(x, y, 0) = f_0(x, y)$ for a given function f_0, and the edges are kept at zero temperature – the solution is then required for $t \geqslant 0$;

(b) the temperature of one side is maintained at a fixed value say $f(x, 0, t) = \phi(x)$, $0 \leqslant x \leqslant a$, while all other sides are kept at zero temperature – in this case a 'steady-state' solution (i.e. one independent of t) is required.

APPENDIX A

Some Results from Elementary Analysis

This appendix, as its name implies, summarizes some of those parts of elementary analysis which are needed for our study of Fourier series and integrals. It is not a simple list of results, some of the definitions and theorems being discussed in detail: however it is emphatically not intended as a systematic treatment of any part of the theory of functions of a single real variable. Results are included or omitted almost always on the grounds that they (or their proofs) have some application to Fourier theory.

A.1 CONTINUITY AND DIFFERENTIATION

We shall consider functions which are defined on intervals of the real line, and which take real or complex values. We use the conventional notation (a, b) for $\{x : a < x < b\}$, the open interval with end points at a, b, and similarly $[a, b] = \{x : a \leqslant x \leqslant b\}$ for the closed interval, with obvious modifications if only one end point is included, or the interval is unbounded in one or both directions.

We begin with the standard definitions of continuity and uniform continuity.

Definition A.1 Let f be a real- or complex-valued function defined on an interval I. We say that f is *continuous at a point a* of I if $f(x) \to f(a)$ through points of I: formally, for each $\varepsilon > 0$, there is a positive δ for which $|f(x) - f(a)| < \varepsilon$, whenever $x \in I$ and $|x - a| < \delta$.

We say that f is *continuous on I* if it is continuous at each point of I; that is, the above definition holds (with a choice of δ which may well vary with the choice of a) at each point of I.

We say that f is *uniformly continuous on I* if the rate at which $f(x) \to f(a)$ as $x \to a$ is independent of the choice of a. More precisely, for each $\varepsilon > 0$, there is a positive δ for which $|f(x) - f(y)| < \varepsilon$ whenever x and y are points of I with $|x - y| < \delta$.

Examples show that a function may be continuous on an interval without being uniformly continuous there. For instance, $f(x) = x^2$ is continuous on $I = \mathbb{R} = (-\infty, \infty)$, and $f(x) = 1/x$ is continuous on $(0, \infty)$ (or $(0, 1)$) but neither is uniformly continuous on the respective intervals. To see this, we take a fixed $h > 0$, and show that $f(x + h) - f(x)$ becomes large, as $x \to \pm \infty$ in the first case, or as $x \to 0$ in the second.

It is of considerable importance in the integration section of this appendix, and elsewhere, to know that this cannot happen if f is defined on a *closed bounded* interval (or indeed on any compact set, for those readers familiar with the more general notion). We sketch the proof of this in a number of stages.

Recall that a sequence of real numbers is said to be *monotone* if it is either increasing ($x_{n+1} \geq x_n$ for all n), or decreasing ($x_{n+1} \leq x_n$ for all n).

Lemma A.2 Let (x_n) be any sequence of real numbers. Then (x_n) contains a monotone sub-sequence.

Proof We say that a point x_m is a *peak point* of the sequence if $x_m \geq x_n$ for all $n \geq m$. (From x_m as a vantage point one can 'see over' all subsequent terms.) Consider the set of all peak points. If this is an infinite set, then the peak points themselves constitute a monotone decreasing sub-sequence of (x_n). It is a finite set, then for some n_1, x_n will not be a peak point if $n \geq n_1$. In particular, x_{n_1} is not a peak point, so let n_2 be the smallest n with $x_{n_2} > x_{n_1}$. Similarly, x_{n_2} is not a peak point, so let n_3 be the smallest n with $x_{n_3} > x_{n_2}$. Continue thus to construct a monotone (strictly) increasing sub-sequence and thus complete the proof. (For those who trouble about such things, this proof does not use the axiom of choice.)

Lemma A.3 Let (x_n) be a bounded sequence of real numbers. Then (x_n) contains a convergent sub-sequence.

Proof By Lemma A.2, (x_n) contains a monotone sub-sequence, which will also be bounded since (x_n) is. But a bounded monotone sequence converges to its supremum if increasing, or to its infimum if decreasing, and the proof is complete.

Lemma A.4 Let f be continuous on the closed bounded interval I. Then f is uniformly continuous on I.

Proof Suppose f is not uniformly continuous on I. Then for some $\varepsilon > 0$, and each $n = 1, 2, 3, \ldots$, there are points x_n, y_n in I, with $|f(x_n) - f(y_n)| \geq \varepsilon$ and $|x_n - y_n| < 1/n$. By Lemma A.3 the sequence (x_n) has a convergent sub-sequence – say $(x_{n_i})_{i=1}^{\infty}$, with $x_{n_i} \to z$ as $i \to \infty$. Since I is closed, z must be a point of I, and since $|x_{n_i} - y_{n_i}| < 1/n$, $y_{n_i} \to z$ also. Then we can find points x_{n_i}, y_{n_i} arbitrarily close to z, with $|f(x_{n_i}) - f(y_{n_i})| \geq \varepsilon$, which is incompatible with the continuity of f at z. This contradiction establishes the result.

We shall take for granted the elementary consequences of continuity, such as the fact that sums, products, and composites of continuous functions are again continuous, and that continuous functions on a closed bounded interval are bounded, attain their bounds (this may be proved in a similar way to Lemma A.4 above) and have the intermediate-value property. Instead we investigate the relationship to a stronger version of continuity which we now define.

Definition A.5 Let α be a real number, $0 < \alpha \leqslant 1$. Let f be defined on an interval I (which may be open or closed, bounded or unbounded). We say that f satisfies a *Lipschitz condition of order α on I* if for some constant M,

$$|f(x) - f(y)| \leqslant M|x - y|^{\alpha}, \text{ for all } x, y \text{ in } I.$$

We shall almost always be concerned with order 1, and simply say that f satisfies a *Lipschitz condition on I* (no mention of order), or is *Lipschitz continuous*, in this case. The example

$$f(x) = \begin{cases} \left\{ \log\left(\dfrac{1}{x}\right) \right\}^{-1}, & 0 < x \leqslant \tfrac{1}{2}, \\ 0, & x = 0 \end{cases}$$

shows that f may be continuous on I (here $[0, \tfrac{1}{2}]$) without satisfying a Lipschitz condition of any order.

If for a fixed x in I and some $M > 0$, $|f(x) - f(a)| \leqslant M|x - a|$, we shall say that f is *Lipschitz continuous at a*. The reader is invited to construct a function which is Lipschitz continuous at one point, but is not continuous (in the sense of Definition A.1) at any other point.

We finish this section with a brief mention of differentiation.

Definition A.6 Let f be defined on an open interval $I = (a, b)$ and let $a < c < b$. We say f is *differentiable at c* if

$$\frac{f(x) - f(c)}{x - c}$$

tends to a limit l, say, as $x \to a$. The limit l is called the *derivative of f at c*, and is denoted $f'(c)$.

We say, f is *differentiable on I* if it is differentiable at each point of I. In this case the function f', which is now defined for all points of I, is called the *derived function* (or just the derivative) of f. (There is an obvious modification of the definition to deal with the end points, when f is defined on a closed interval.)

Well-known examples show that a continuous function need not be differentiable at an individual point, or indeed at any point at all, and that a derived function f' when it exists need not be continuous everywhere. (A derived function must have *some* points of continuity, but a proof of this would take us too far afield.)

We shall assume the elementary properties of differentiable functions, including the product and chain rules for differentiation and the mean-value theorem.

Lemma A.7 Let f be differentiable on an interval I, and suppose f' is bounded: for some M, $|f'(x)| \leqslant M$ for all x in I. Then f satisfies a Lipschitz condition on I.

Proof Let x, y be points of I. By the mean-value theorem, there is a point z between x and y, for which

$$f(x) - f(y) = (x - y)f'(z),$$

and the result is immediate from this.

The converse of this result, namely that a Lipschitz function must be differentiable 'almost everywhere' is true, but beyond our reach.

A.2 INTEGRATION

This section outlines the Riemann process for integrating a function on a bounded interval, and some of its principal properties.

Definition A.8 Let $I = [a, b]$ be a closed bounded interval. A *dissection* (or partition) of I is a finite set $D = \{x_0, x_1, x_2, \ldots, x_n\}$ of points of I for which

$$a = x_0 < x_1 < \cdots < x_{n-1} < x_n = b.$$

For instance, $\{0, \frac{1}{3}, 1\}$ and $\{0, \frac{1}{10}, \frac{2}{10}, \frac{3}{10}, \ldots, \frac{9}{10}, 1\}$ are dissections of $[0, 1]$. The length of the longest subinterval $[x_{i-1}, x_i]$ $(i = 1, 2, \ldots, n)$ defined by D is called the *norm* of D: $\|D\| = \max\{x_i - x_{i-1}; \ 1 \leqslant i \leqslant n\}$. Thus $\|D\| = \frac{2}{3}$, $\frac{1}{10}$ respectively in the examples given.

Now suppose f is a real-valued function defined on $[a, b]$. Throughout the whole of this section we shall suppose that f is bounded: for some real numbers m and M, we have

$$m \leqslant f(x) \leqslant M \text{ for all } x \text{ in } [a, b].$$

Suppose that a dissection D is given. Then for each $i = 1, 2, \ldots, n$, let m_i, M_i be respectively the infimum and supremum of f on $[x_{i-1}, x_i]$, and let y_i be any point of $[x_{i-1}, x_i]$. Then we obviously have $m \leqslant m_i \leqslant f(y_i) \leqslant M_i \leqslant M$, and hence

$$m(b-a) \leqslant \sum_{i=1}^{n} m_i(x_i - x_{i-1}) \leqslant \sum_{i=1}^{n} f(y_i)(x_i - x_{i-1}) \leqslant \sum_{i=1}^{n} M_i(x_i - x_{i-1}) \leqslant M(b-a)$$

The three sums appearing here will be denoted

$$S_L(f, D), \quad S_y(f, D), \quad S_U(f, D)$$

respectively, where the L and U denote 'lower' and 'upper', and the suffix y

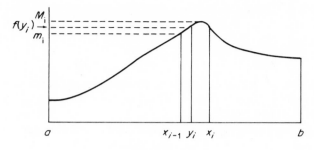

Fig. A-1.

reminds us of the dependence on the choice of points (y_i). Each of the sums gives an approximation to the geometrical notion of the area under the curve; see Figure A.1

In particular, if f is negative on any subinterval then m_i and M_i will be negative and there will be a negative contribution to the corresponding sum.

These sums enable us to define the integrability of f.

Definition A.9 Let f be a real-valued bounded function on $[a, b]$ and let sums S_L, S_U be defined as above for each dissection D of $[a, b]$. The numbers I_L, I_U defined by

$$I_L = I_L(f) = \sup_D S_L(f, D),$$

$$I_U = I_U(f) = \inf_D S_U(f, D),$$

are called the *lower* and *upper integrals of f over* $[a, b]$, and are also denoted $\int_a^b f(x)\, dx$, $\overline{\int_a^b} f(x)\, dx$ respectively. It is important to realize that I_L and I_U are defined for all bounded f whether continuous or not.

We say that f is *integrable over* $[a, b]$ if $I_L = I_U$. The common value is called the *integral of f over* $[a, b]$ and is denoted $\int_a^b f(x)\, dx$.

It is a straightforward but slightly messy exercise (which we shall omit) to show that this definition is equivalent to the following:

Definition A.9′ Let f be a real-valued function on $[a, b]$, and let $S_y(f, D) = \sum_{i=1}^n f(y_i)(x_i - x_{i-1})$, for a given dissection D and choice of points (y_i), as described above. Then f is *integrable over* $[a, b]$ if $S_y(f, D)$ tends to a limit L as $\|D\| \to 0$: more precisely, for each $\varepsilon > 0$, there is a $\delta > 0$, such that $|S_y(f, D) - L| < \varepsilon$ for every D with $\|D\| < \delta$, and every choice of (y_i) relative to D.

The number L is the *integral* of f over $[a, b]$ and is the same as the common value of I_L and I_U in Definition A.9.

From now on we shall feel free to use either of these definitions, as appropriate. In particular, on taking a dissection of $[a, b]$ into n equal intervals, by $x_i = a + i(b - a)/n$, and $y_i = x_i$, $i = 0, 1, 2, \ldots, n$, we obtain the following lemma.

Lemma A.10 Let f be a real-valued integrable function on $[a,b]$.
Then

$$\frac{b-a}{n}\sum_{i=1}^{n} f\left(a+i\frac{b-a}{n}\right) \to \int_a^b f(x)\,dx \quad \text{as} \quad n\to\infty.$$

The integration process is extended to complex-valued functions by writing $f=f_1+if_2$ where f_1 and f_2 are real valued, and defining $\int_a^b f(x)\,dx$ to be $\int_a^b f_1(x)\,dx + i\int_a^b f_2(x)\,dx$.

We shall not attempt a complete description of the class of integrable functions: however, for our purposes it is important that all FC-functions are integrable, and we prove this next.

Theorem A.11 Let f be an FC-function on $[a,b]$. Then f is integrable over $[a,b]$.

Proof For a given dissection D of $[a,b]$, we have

$$S_U(f,D) - S_L(f,D) = \sum_{i=1}^{n}(M_i - m_i)(x_i - x_{i-1}),$$

and Definition A.9 tells us that to show that f is integrable we must find some D with $S_U - S_L$ small.

Suppose $\varepsilon > 0$ is given. Suppose that f has discontinuities at points z_1, $z_2,\ldots,z_N$ and that $|f(x)| \leq M$ for all x in $[a,b]$. Construct disjoint open intervals $I_1, I_2,\ldots,I_N$ about each point $z_1,\ldots,z_N$, each of length at most $\varepsilon/4MN$. Let $I = I_1\cup I_2\cup\cdots\cup I_N$, and let $S = [a,b]\backslash I$. Then S will be a union of closed bounded intervals, some of which may reduce to a single point, on each of which f is continuous, so that from Lemma A.4 we may deduce that f is uniformly continuous on S. Hence for some $\delta > 0$, we have $|f(x)-f(y)| < \varepsilon/2(b-a)$ if x, y are in S and $|x-y| < \delta$.

Now construct a dissection of $[a,b]$ using the intervals $I_1, I_2,\ldots,I_N$ together with any dissection of S into intervals of length less than δ.

The intervals $I_1,\ldots,I_N$ contribute at most

$$N2M\frac{\varepsilon}{4MN} = \frac{\varepsilon}{2}\text{ to the sum } S_U - S_L,$$

since $M_i - m_i \leq 2M$ for any possible subinterval of $[a,b]$. The intervals which make up S contribute at most

$$\frac{\varepsilon}{2(b-a)}(b-a) = \frac{\varepsilon}{2}\text{ to } S_U - S_L,$$

since their total length is no more than $b-a$.

Hence we have found a D with $S_U - S_L < \varepsilon$, and so f is integrable.

The *fundamental theorem of calculus* takes the following form.

Theorem A.12 Let f be differentiable on the interval $[a,b]$, and let f' be an FC-function. Then

$$\int_a^b f'(x)\,dx = f(b) - f(a).$$

Proof For an arbitrary dissection $D = \{x_0, x_1, \ldots, x_n\}$ we may use the mean-value theorem to write

$$f(b) - f(a) = \sum_{i=1}^n f(x_i) - f(x_{i-1}) = \sum_{i=1}^n (x_i - x_{i-1})f'(y_i)$$

where the points y_i are in the intervals (x_{i-1}, x_i). But by Definition A.9' this sum tends to $\int_a^b f'(x)\,dx$ as $\|D\| \to 0$, and the result follows. (The example $f(x) = x^2 \sin(1/x)$ if $x \neq 0$, $f(0) = 0$, shows that a function can have a derivative which is an FC-function which is not continuous.)

Once again we shall assume the elementary properties of the integration process, such as its linearity, additivity with respect to intervals, estimation properties

$$\left(\left| \int_a^b f(x)\,dx \right| \leqslant \int_a^b |f(x)|\,dx \right),$$

etc., and the rules for integration by parts and substitution. We indicate briefly how the integration process is extended to functions which are defined on the whole of the real line.

Definition A.13 Let f be defined for all real numbers, and suppose that
(i) f is an FC-function on *every* bounded interval $[a, b]$; and
(ii) for some M, $\int_a^b |f(x)|\,dx \leqslant M$ for all a, b.
   Then we say that f is an *FV-function* on the real line. The fact that $\int_a^b |f(x)|\,dx$ increases as $a \to -\infty$, $b \to +\infty$, together with (ii) shows that $\int_a^b |f(x)|\,dx$ approaches a limit as $a \to -\infty$, $b \to +\infty$ independently. Since for any a', b',

$$\left| \int_{a'}^{b'} f(x)\,dx \right| \leqslant \int_{a'}^{b'} |f(x)|\,dx,$$

the general principle of convergence (cf. Theorem A.20 below) shows that $\int_a^b f(x)\,dx$ also approaches a limit as $a \to -\infty, b \to +\infty$: this limit will be denoted $\int_{-\infty}^\infty f(x)\,dx$; it exists for all FV-functions.
   From now on we shall write $\mathbb{R}$ for the set of all real numbers, and write 'f is defined on $\mathbb{R}$' instead of the more cumbersome 'f is defined for all real numbers'.

Example A.14 (i) For any integer n let $f(x) = 1$ on intervals of the form

164

$(2n, 2n + 1)$, $f(x) = -1$ elsewhere. Then f is an FC-function on each bounded interval $[a, b]$, but $|f(x)| = 1$ for all x, so $\int_a^b |f(x)| \, dx = b - a$ is not bounded, condition (ii) of Definition A.13 is violated, and f is not an FV-function.

(ii) Let $f(x) = 1/(1 + x^2)$ for all real x.

Then from elementary calculus we know that

$$\int_a^b \frac{dx}{1 + x^2} = \tan^{-1} b - \tan^{-1} a < \frac{\pi}{2} - \left(-\frac{\pi}{2} \right) = \pi.$$

Hence f is an FV-function (with no discontinuities), and evidently $\int_{-\infty}^{\infty} (1 + x^2)^{-1} \, dx = \pi$.

(iii) The reader is invited to construct an example to show that a continuous FV-function need not tend to zero as $x \to \pm \infty$; indeed it need not even be bounded on the whole of $\mathbb{R}$. (Construct a function f which is positive and has for each integer $n \neq 0$, $f(n) = n$, $\int_{n-1/2}^{n+1/2} f(x) \, dx < 1/n^2$.)

(iv) For positive integer values of n, let $f(x) = (-1)^{n-1}/n$ on the interval $(n - 1, n)$, $f(x) = 0$ elsewhere.

Then f is an FC-function on any subinterval of $\mathbb{R}$, and the integral $\int_0^b f(x) \, dx$ approaches a limit as $b \to \infty$. (This statement is equivalent to the convergence of the series $\sum_1^{\infty} (-1)^{n-1}/n$.)

However, $\int_0^b |f(x)| \, dx$ is not bounded (the series $\sum_1^{\infty} 1/n$ diverges), so that f is not an FV-function. Another example of this appears in theorem B.5(ii).

For functions which are defined on $\mathbb{R}$, including those which are defined on an interval and extended to $\mathbb{R}$ by the requirement of periodicity (see Section 1.2), we have the operation of translation which consists of taking the graph of f and moving it to the right or left. More formally, we make the following definition.

Definition A.15 Let f be defined on $\mathbb{R}$ and let h be any real number. Then the *translate of f by h* is the function f_h defined by $f_h(x) = f(x - h)$ for all real x.

Figure A.2 illustrates this with a positive value of h.

We can measure the distance from f to f_h in any of the ways discussed in Section 1.4: in particular, the distance $d_1(f, f_h)$ is needed for Lemma 1.15.

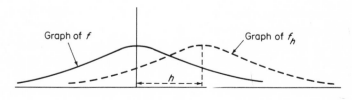

Graph of f Graph of f_h

h

Fig. A-2.

Definition A.16 The distance

$$d_1(f, f_h) = \int_{-\infty}^{\infty} |f(x) - f_h(x)| \, dx$$

$$= \int_{-\infty}^{\infty} |f(x) - f(x - h)| \, dx$$

is called the *integral modulus of continuity* of f, and will be denoted $\omega_f(h)$. Notice that $\omega_f(h) = \omega_f(-h)$, so we may assume $h > 0$ if needed. In the case of a periodic function the above integral may not converge, so we replace the above definition by

$$\omega_f(h) = \int_a^{a+p} |f(x) - f(x - h)| \, dx$$

where a is any real number and p the period of f.

(The distance $d_0(f, f_h) = \sup_x |f(x) - f_h(x)|$ is called the *modulus of continuity* of f, but we shall make rather less use of this concept. Notice that the Lipschitz continuity of f (Definition A.5) may be restated as $d_0(f, f_h) \leqslant M|h|$, while the uniform continuity (Definition A.1) simply states that $d_0(f, f_h) \to 0$ as $h \to 0$.)

The fact that $d_1(f, f_h) = \omega_f(h) \to 0$ as $h \to 0$ for any FV-function (or periodic FC-function) is important: the proof follows from the following lemma on approximation by continuous functions, which is also a useful result in its own right.

Lemma A.17 (i) Let f be an FC-function on $[a, b]$. Then for each $\varepsilon > 0$, there exists a continuous function g on $[a, b]$, such that $\int_a^b |f(x) - g(x)| \, dx < \varepsilon$.

(ii) Let f be an FV-function on $\mathbb{R}$. Then for each $\varepsilon > 0$, there exists a continuous function g on $\mathbb{R}$, which vanishes for all x not in some bounded interval, such that $\int_{-\infty}^{\infty} |f(x) - g(x)| \, dx < \varepsilon$.

Proof We first reduce the proof of (ii) to that of (i), then prove (i).

Suppose $\varepsilon > 0$ is given. From Definition A.13 we deduce that for some interval $I = [a, b]$ say, $\int_{I'} |f(x)| \, dx < \frac{1}{2}\varepsilon$.

(Here $I' = \mathbb{R} \backslash I = (-\infty, a) \cup (b, \infty)$). Hence if we can construct a continuous g on $[a, b]$ with $g(a) = g(b) = 0$ and $\int_a^b |f(x) - g(x)| \, dx < \frac{1}{2}\varepsilon$, then this g extends to a function on $\mathbb{R}$ (by defining $g(x) = 0$ if x is not in I) which satisfies (ii). Thus it is sufficient to prove (i) with the additional restriction that $g(a) = g(b) = 0$.

Suppose then that f has discontinuities at $x_1, x_2, \ldots, x_N$ in $[a, b]$, and that $|f(x)| \leqslant M$ for all x in $[a, b]$. Let $t = \varepsilon/2M(N + 2)$. Construct disjoint open intervals $I_0, I_1, I_2, \ldots, I_N, I_{N+1}$, containing $a, x_1, x_2, \ldots, x_N, b$ respectively, each of length at most t. Let

$$S = [a, b] \backslash \left(\bigcup_{n=0}^{N+1} I_n \right).$$

Then S is a union of closed intervals (some of which may reduce to a single point) on each of which f is continuous.

Define $g(x) = f(x)$ if x is in S, and $g(a) = g(b) = 0$. This definition includes the end points of each interval I_n, and hence g may be extended to I_n as a linear function which interpolates these values. This ensures both that g is continuous and that $|g(x)| \leqslant M$ on the whole of $[a, b]$. Then

$$\int_a^b |f(x) - g(x)| \, dx = \sum_{n=0}^{N+1} \int_{I_n} |f(x) - g(x)| \, dx$$
$$\leqslant 2M(\text{sum of lengths of } I_n)$$
$$\leqslant 2M(N + 2)t = \varepsilon, \text{ as required.}$$

Corollary A.18 Let f be an FV-function on $\mathbb{R}$, or an FC-function on $[a, b]$. Then $\omega_f(h) \to 0$ as $h \to 0$.

Proof In either case, given $\varepsilon > 0$, we choose a continuous g such that $\int |f(x) - g(x)| \, dx < \frac{1}{3}\varepsilon$. (The $\int$ sign denotes either $\int_{-\infty}^{\infty}$ or $\int_a^b$ corresponding to the two cases.) In either case choose g to be continuous and zero except on a closed bounded interval $[a, b]$; then by Lemma A.4 we can find $\delta > 0$ such that $|g(x) - g(y)| < \varepsilon/3(b - a)$ whenever $|x - y| < \delta$. Then if $|h| < \delta$, $\int |g(x) - g_h(x)| \, dx < \varepsilon/3$, and hence

$$\int |f(x) - f_h(x)| \, dx \leqslant \int |f(x) - g(x)| \, dx + \int |g(x) - g_h(x)| \, dx + \int |g_h(x) - f_h(x)| \, dx$$

$$< 3\frac{\varepsilon}{3} = \varepsilon, \qquad \text{as required.}$$

We finish this section with a proof of the classic Cauchy–Schwarz inequality.

Theorem A.19 (i) Let (a_n), (b_n) be any sequences of complex numbers. Then

$$\left| \sum_n a_n b_n \right|^2 \leqslant \sum_n |a_n|^2 \sum_n |b_n|^2,$$

in the sense that if both the series on the right converge, so does the series on the left, and the stated inequality is valid. There is equality if and only if for some complex numbers α, β, $\alpha a_n = \beta \bar{b}_n$ for all n.

(ii) Let f, g be FC-functions on $[a, b]$. Then

$$\left| \int_a^b f(x)g(x) \, dx \right|^2 \leqslant \int_a^b |f(x)|^2 \, dx \int_a^b |g(x)|^2 \, dx.$$

There is equality if and only if for some complex numbers α, β, the functions αf and $\beta \bar{g}$ are almost equal (Definition 1.7).

Proof We prove (ii), leaving the reader to supply the analogous argument to prove (i), and also to frame and prove an extension applicable to FV-functions on $\mathbb{R}$. Suppose then that f, g are FC-functions on $[a, b]$ and that λ is any complex number. Evidently,

$$\int_a^b |f(x) + \lambda \bar{g}(x)|^2 \, dx \geqslant 0,$$

i.e. $\displaystyle\int_a^b (f(x) + \lambda \bar{g}(x))(\bar{f}(x) + \bar{\lambda} g(x)) \, dx$

$$= \int_a^b |f(x)|^2 \, dx + \lambda \int_a^b \bar{f}(x) \bar{g}(x) \, dx + \bar{\lambda} \int_a^b f(x) g(x) \, dx$$

$$+ |\lambda|^2 \int_a^b |g(x)|^2 \, dx \geqslant 0.$$

Now if $\int_a^b |g(x)|^2 \, dx = 0$, then g must be almost zero (Definition 1.7 again), so that fg is almost zero, and the result is true with both sides $= 0$. If not, choose

$$\lambda = - \int_a^b f(x) g(x) \, dx \Big/ \int_a^b |g(x)|^2 \, dx.$$

This gives

$$\int_a^b |f(x)|^2 \, dx - \left| \int_a^b f(x) g(x) \, dx \right|^2 \Big/ \int_a^b |g(x)|^2 \, dx \geqslant 0$$

which is the required inequality. There is equality if and only if either g is almost zero, or, for the chosen value of λ, $f + \lambda \bar{g}$ is almost zero, which is equivalent to the given condition.

[In the notation of inner products (Definition 1.3), this result simply states that $|(f, g)|^2 \leqslant (f, f)(g, g)$.]

A.3 UNIFORM CONVERGENCE

This section outlines the results which we need on sequences of functions, relating to the continuity, differentiability, and integrability of the limit function. We begin with a brief mention of the *general principle of convergence* (or Cauchy criterion) for sequences of complex numbers.

Theorem A.20 Let $(z_n)_{n=1}^{\infty}$ be a sequence of real or complex numbers. Then $(z_n)_1^{\infty}$ is convergent if and only if for each $\varepsilon > 0$, there is a value of $n = n_0$, say, such that

$$|z_n - z_m| < \varepsilon \text{ whenever } n, m \text{ are both } \geqslant n_0.$$

Proof If $z_n \to l$ as $n \to \infty$, then given $\varepsilon > 0$, there is an n_0 such that $|z_n - l| < \frac{1}{2}\varepsilon$ if $n \geqslant n_0$. Then

$$|z_n - z_m| < |z_n - l| + |z_m - l| < \tfrac{1}{2}\varepsilon + \tfrac{1}{2}\varepsilon = \varepsilon \text{ if } m, n \geqslant n_0.$$

Conversely, if $|z_n - z_m| \to 0$ as $m, n \to \infty$ (which is what the given condition says), then obviously the same condition holds for the real and imaginary parts of the sequence, so that there is no loss of generality in assuming the sequence to be real valued. Also taking $\varepsilon = 1$ in the definition shows that for some n_1, $|z_n - z_{n_1}| < 1$ if $n \geqslant n_1$, so for *all* n,

$$|z_n| \leqslant \max \{|z_1|, |z_2|, \dots, |z_{n_1 - 1}|, |z_{n_1}| + 1\}$$

and hence the sequence is bounded.

It follows from Lemma A.3 that there is a sub-sequence of (z_n) which is convergent to a limit l. Hence given $\varepsilon > 0$ we know both:

(a) for some n_0, $|z_n - z_m| < \varepsilon$ if $m, n \geq n_0$; and
(b) for some $m \geq n_0$, $|z_m - l| < \varepsilon$, by choosing z_m in the sub-sequence which converges to l.

Combining these we see that $|z_n - l| < 2\varepsilon$ if $n \geq n_0$, and thus the whole sequence is convergent, as required.

A corollary of this is the elementary result for series that if $\sum_n |a_n|$ is convergent (i.e. $\sum_n a_n$ is absolutely convergent) then $\sum_n a_n$ is convergent: this follows on applying Theorem A.20 to the partial sums of the series.

We can now move on to our main concern, which is the convergence of sequences or series of functions.

Definition A.21 Let $(f_n)_{n=1}^{\infty}$ be a sequence of functions defined on an interval I (I may be open or closed, bounded or unbounded). We say that (f_n) *converges pointwise on I* if there is a function f on I such that $f_n(x) \to f(x)$ as $n \to \infty$, for each x in I. More formally, for each x in I, and each $\varepsilon > 0$, there is an n_0 such that $|f_n(x) - f(x)| < \varepsilon$ if $n \geq n_0$. Notice that n_0 will depend in general on ε and x.

We say that (f_n) *converges uniformly on I* if the rate at which $f_n(x)$ approaches $f(x)$ is independent of the point x (compare Definition A.1) More formally, for each $\varepsilon > 0$, there is an n_0 (depending on ε alone) such that $|f_n(x) - f(x)| < \varepsilon$ for all $n \geq n_0$ and all x in I.

We can obtain corresponding definitions for the pointwise and uniform convergence of *series* of functions by applying this definition to their partial sums.

The general principle of convergence (Theorem A.20) extends to uniform convergence in the following way: the sequence (f_n) is uniformly convergent on the interval I if and only if for each $\varepsilon > 0$, there is an n_0 such that $|f_n(x) - f_m(x)| < \varepsilon$ for all $m, n \geq n_0$, and x in I.

Examples A.22 (i) Let (f_n) be defined on $[0, 1]$ by

$$f_n(x) = \begin{cases} 1 - nx, & 0 \leq x \leq 1/n \\ 0, & 1/n \leq x \leq 1 \end{cases} \qquad \text{(Fig. A.3)}$$

Here we have $f_n(0) = 1$ for all n, while if $x > 0$, and $1/n < x$, then $f_n(x) = 0$. Hence for each x in $[0, 1]$, $f_n(x) \to f(x)$ as $n \to \infty$, where $f(0) = 1$, while $f(x) = 0$ if $x > 0$. In this example we have a sequence of continuous functions which converges pointwise to a discontinuous limit.

(ii) Let (f_n) be defined on $\mathbb{R}$ by $f_n(x) = \sqrt{x^2 + 1/n}$ (positive square root), as in Fig. A.4.

Here $f_n(x) \to \sqrt{x^2} = |x|$ as $n \to \infty$, and the convergence is uniform on $\mathbb{R}$, since

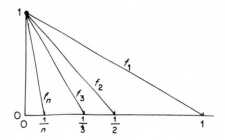

Fig. A-3.

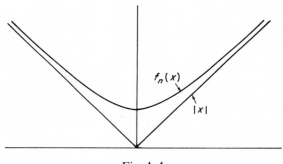

Fig. A-4.

the estimate $0 < f_n(x) - |x| \leqslant 1/\sqrt{n}$ is independent of x. In this example we have a sequence of differentiable functions which converge uniformly to a non-differentiable limit.

A particularly easy way of obtaining uniformly convergent series is by means of the following useful result.

Theorem A.23 (Weierstrass M-test) Let (f_n) be a sequence of functions defined on an interval I, and suppose that for each n, we have an estimate of the form $|f_n(x)| \leqslant M_n$ for all x in I, where $\sum_n M_n$ is convergent. Then $\sum_n f_n(x)$ is uniformly convergent.

Proof Since $\sum_n M_n$ is convergent, it follows that $\sum_n |f_n(x)|$ is convergent for each x in I, and hence that $\sum_n(x)$ is pointwise convergent (see the remark following Theorem A.20). Let $f(x) = \sum_n f_n(x)$ for all x in I: we have to show the convergence is uniform.

Given $\varepsilon > 0$, choose N such that $\sum_N^\infty M_n < \varepsilon$.

Then if $n \geqslant N, |f(x) - \sum_{r=1}^n f_r(x)| \leqslant \sum_{r=n+1}^\infty |f_r(x)| \leqslant \sum_{r=n+1}^\infty M_r < \varepsilon$, so that $\sum_{r=1}^n f_r(x) \rightarrow f(x)$ uniformly as required.

Corollary A.24 Let (a_n) be a positive sequence for which $\sum_n a_n$ is convergent, and let (f_n) be a uniformly bounded sequence of functions on an interval I: for some $M > 0$, $|f_n(x)| \leqslant M$ for all n and all x in I.
   Then $\sum_n a_n f_n(x)$ is uniformly convergent on I.

Proof Apply Theorem A.23 with $M_n = M a_n$.
   For example, we may take $f_n(x) = \cos nx$, $\sin nx$, or e^{inx} in Corollary A.24, to obtain the uniform convergence of such series as

$$\sum_1^\infty \frac{1}{n^2} \cos nx, \quad \sum_{-\infty}^\infty 2^{-|n|} e^{inx}, \text{ etc.}$$

We now suppose that we have a sequence (f_n) of continuous functions which converges to a limit function f on I. Example A22(i) shows that f may fail to be continuous, but our next result shows that this is impossible if the convergence is uniform.

Theorem A.25 Let (f_n) be a sequence of functions which converges uniformly on an interval I to a limit function f. Then
   (i) if for some x_0 in I, each f_n is continuous at x_0, then so is f; and
   (ii) if each f_n is continuous on I, then so is f.

Proof (ii) is an immediate consequence of (i), so we prove (i) only.
   Suppose $\varepsilon > 0$ is given; choose n_0 such that $|f_n(x) - f(x)| < \frac{1}{3}\varepsilon$ for all $n \geqslant n_0$ and x in I. We know that f_{n_0} is continuous at x_0, and hence for some $\delta > 0$, $|f_{n_0}(x_0) - f_{n_0}(x)| < \frac{1}{3}\varepsilon$ whenever $|x - x_0| < \delta$.
   We then have

$$|f(x_0) - f(x)| \leqslant |f(x_0) - f_{n_0}(x_0)| + |f_{n_0}(x_0) - f_{n_0}(x)| + |f_{n_0}(x) - f(x)|$$
$$< 3 \cdot \tfrac{1}{3}\varepsilon = \varepsilon, \quad \text{if} \quad |x - x_0| < \delta,$$

and hence f is continuous at x_0, as required.

   There is a partial converse of this result which will be useful to us on several occasions, which states that *monotone* pointwise convergence of a sequence of continuous functions to a continuous limit must be uniform.

Theorem A.26 (Dini) Let (f_n) be a sequence of continuous real-valued functions which converges monotonically to a continuous limit f, at each point of a closed bounded interval I. Then $f_n \to f$ uniformly on I.

Proof We may suppose for simplicity that $f(x) = 0$ on I: if not, consider the sequence $(f_n - f)$. Suppose the conclusion fails. Then for some positive $\varepsilon = \varepsilon_1$ say, and each $n = 1, 2, 3, \ldots$, there is some x_n in I with $|f_n(x_n)| \geqslant \varepsilon_1$. By Lemma A.3, there is a sub-sequence of (x_n) which converges to some point, say t, of I. But we know that $f_n(t) \to 0$ as $n \to \infty$, so there is some $n = n_1$ say with $|f_{n_1}(t)| < \frac{1}{2}\varepsilon_1$.

Also f_{n_1} is continuous at t, so for some $\delta > 0$,

$$|f_{n_1}(x) - f_{n_1}(t)| < \tfrac{1}{2}\varepsilon_1 \quad \text{if} \quad |x - t| < \delta,$$

and hence $|f_{n_1}(x)| < \tfrac{1}{2}\varepsilon_1 + \tfrac{1}{2}\varepsilon_1 = \varepsilon_1$ if $|x - t| < \delta$.

But for each x, the convergence of $f_n(x)$ to 0 is monotone, so that for all $n \geqslant n_1$ we again have $|f_n(x)| < \varepsilon_1$ if $|x - t| < \delta$. However, there are some points x_n with $|x_n - t| < \delta$, since a sub-sequence of (x_n) converges to t, and this leads us to the contradictory result that $f_n(x_n) < \varepsilon_1$ for these x_n, and thus the result is established.

(Notice that the convergence may be monotone increasing at some points of I and decreasing at others. For a neat application of this result, see Theorem B.5 (i)).

We next consider the relation between uniform convergence and integration. For most purposes the following result (which is far from best possible) will suffice.

Theorem A.27 Let (f_n) be a sequence of FC-functions on the interval $[a,b]$ which converges uniformly on $[a,b]$ to an FC-function f.

Then $$\int_a^b f_n(x)dx \to \int_a^b f(x)dx \text{ as } n \to \infty.$$

Proof Given $\varepsilon > 0$, choose n_0 such that $|f_n(x) - f(x)| < \varepsilon/(b-a)$ for all $n \geqslant n_0$ and x in $[a,b]$.

Then $$\left| \int_a^b f_n(x)\,dx - \int_a^b f(x)\,dx \right| \leqslant \int_a^b |f_n(x) - f(x)|\,dx \leqslant \varepsilon$$

$$\text{if} \quad n \geqslant n_0,$$

and the result follows.

Note It is not generally true that a uniform limit of FC-functions is again an FC-function (the reader is invited to construct an example) – hence the need for the hypothesis on f. A more sophisticated approach to integration would deduce the integrability of f from the other hypotheses.

It is widely known that Theorem A27 remains valid if we replace 'uniformly convergent' by 'pointwise convergent + uniformly bounded' – it is also widely believed that the proof is exceptionally difficult; we shall give a relatively easy proof as Theorem A.29 below. The following examples illustrate the need for some restrictive hypotheses.

Examples A.28 (i) Let

$$f_n(x) = \begin{cases} n^2 x & \text{for } 0 \leqslant x \leqslant 1/n, \\ 2n - n^2 x & \text{for } 1/n < x \leqslant 2/n, \\ 0 & \text{elsewhere on } [0,1], \end{cases}$$

for $n = 2, 3, \ldots$.

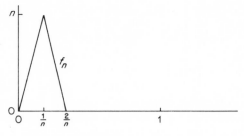

Fig. A-5.

Then each f_n is continuous and $f_n(x) \to 0$ for each x in $[0,1]$, but one easily verifies that $\int_0^1 f_n(x)\mathrm{d}x = 1$ for all n, so that $\int_0^1 f_n(x)\mathrm{d}x$ does not tend to zero. We may see by considering (nf_n) that a sequence may tend to zero while its integrals tend to infinity. In this example the convergence is of course not uniform (nor is it uniformly bounded – see Theorem A.29).

(ii) Let

$$f(x) = \begin{cases} \dfrac{1}{n}\left(1 - \dfrac{x}{n}\right) & \text{of } 0 \leqslant x \leqslant n \\ 0 & \text{if } x > n \end{cases} \qquad \text{(Fig. A·6)}.$$

f_n is continuous and tends to zero uniformly on $\mathbb{R}$, but $\int_0^\infty f_n(x)\mathrm{d}x = 1$ and so does not tend to zero. Thus Theorem A.27 is essentially limited to bounded intervals.

Theorem A.29 Let (f_n) be a sequence of FC-functions on an interval $[a,b]$ which converges pointwise to an FC-function f on $[a,b]$. Suppose also that the sequence is uniformly bounded: for some M we have

$$|f_n(x)| \leqslant M \text{ for all } n = 1,2,3,\ldots \text{ and } x \text{ in } [a,b].$$

Then $\int_a^b f_n(x)\mathrm{d}x \to \int_a^b f(x)\mathrm{d}x$ as $n \to \infty$.

Proof We suppose without loss of generality that $f_n(x) \to 0$ for each x in $[a,b]$, and prove that $\int_a^b f_n(x)\mathrm{d}x \to 0$. For each $n = 1,2,3,\ldots$, define functions g_n, h_n on $[a,b]$ by

$$h_n(x) = \sup\{f_m(x); m \geqslant n\},$$
$$g_n(x) = \inf\{f_m(x); m \geqslant n\}.$$

Fig. A-6.

Then the sequences (h_n) and (g_n) are uniformly bounded and $g_n(x) \leqslant 0 \leqslant h_n(x)$ for all n and x. Also (g_n) increases monotonically and (h_n) decreases monotonically, both with limit zero. Notice that h_n and g_n will *not* generally be FC-functions: however, the definition of the integration process (Definition A.9) shows that we have

$$\overline{\int}_a^b g_n(x)\mathrm{d}x \leqslant \int_a^b f_n(x)\mathrm{d}x \leqslant \overline{\int}_a^b h_n(x)\mathrm{d}x, \quad \text{for each } n.$$

Hence the result will be established if we can prove the following lemma concerning sequences (h_n) which decrease to zero. (The corresponding result for (g_n) follows on applying Lemma A.30 to $(-g_n)$).

Lemma A.30 Let (h_n) be a uniformly bounded sequence of functions which is monotone decreasing with limit zero at each point of $[a,b]$.

Then $$\overline{\int}_a^b h_n(x)\mathrm{d}x \to 0 \quad \text{as} \quad n \to \infty.$$

Proof (Adapted from 27.17 of Jameson, 1974)

Suppose $\varepsilon > 0$ is given. For an arbitrary positive bounded function h, Definition A.9 shows that there is a dissection D of $[a,b]$ such that

$$S_L(h,D) > \overline{\int}_a^b h(x)\mathrm{d}x - \varepsilon.$$

Here $S_L(h,D) = \sum_{i=1}^n m_i(x_i - x_{i-1})$, with the notations established following Definition A.8, and on defining a function s on $[a,b]$ by

$$\left. \begin{array}{l} s(x_i) = 0 \\ s(x) = m_i \quad \text{if} \quad x_{i-1} < x < x_i \end{array} \right\} \quad i = 1,2,\ldots,n,$$

we see that $S_L(h,D) = \int_a^b s(x)\mathrm{d}x$. By modifying s to be a linear function in the neighbourhood of each x_i, as in Fig. A.7, we see that for any h, and $\varepsilon > 0$, we can find a *continuous* positive function s on $[a,b]$ such that $\int_a^b s(x)\mathrm{d}x > \overline{\int}_a^b h(x)\mathrm{d}x - \varepsilon$.

We apply this construction to each of the functions h_n: in fact for the given $\varepsilon > 0$ and $n = 1,2,\ldots$ we choose a continuous positive function s_n on $[a,b]$ with $0 \leqslant s_n \leqslant h_n$, and

$$\int_a^b s_n(x)\,\mathrm{d}x > \overline{\int}_a^b h_n(x)\,\mathrm{d}x - \varepsilon/2^n.$$

Now let $t_n = \min(s_1, s_2, \ldots, s_n)$: the sequence t_n is thus a decreasing sequence of positive continuous functions on $[a,b]$. Since $0 \leqslant t_n \leqslant s_n \leqslant h_n$, we deduce that $t_n \to 0$ pointwise on I, and hence also uniformly by Dini's Theorem, A.26. By Theorem A.27 it now follows that $\int_a^b t_n(x)\,\mathrm{d}x \to 0$ as $n \to \infty$, and hence that for some N, $\int_a^b t_N(x)\mathrm{d}x < \varepsilon$. We shall show that $\overline{\int}_a^b h_N(x)\mathrm{d}x < 2\varepsilon$, and the result will be proved.

174

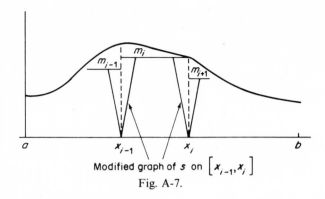

Modified graph of s on $\left[x_{i-1}, x_i\right]$

Fig. A-7.

We know firstly that $t_1 = s_1$, so that

$$\int_a^b t_1(x)\, dx = \int_a^b s_1(x)\, dx > \int_a^b h_1(x)\, dx - \tfrac{1}{2}\varepsilon.$$

From now on we shall write, as is conventional, $f \wedge g$, $f \vee g$ for $\min(f, g)$, $\max(f, g)$ respectively. In this notation

$$t_2 = s_1 \wedge s_2 = t_1 \wedge s_2, \text{ so that}$$
$$t_2 = t_1 + s_2 - t_1 \vee s_2, \text{ and}$$

$$\int_a^b t_2(x)\, dx = \int_a^b t_1(x)\, dx + \int_a^b s_2(x)\, dx - \int_a^b (t_1 \vee s_2)(x)\, dx$$

$$> \int_a^b h_1(x)\, dx - \tfrac{1}{2}\varepsilon + \int_a^b h_2(x)\, dx - \tfrac{1}{4}\varepsilon - \int_a^b h_1(x)\, dx,$$

where in the last term we have used the fact that $t_1 \vee s_2 \leqslant h_1 \vee h_2 = h_1$.

Hence
$$\int_a^b t_2(x)\, dx > \int_a^b h_2(x)\, dx - (\tfrac{1}{2} + \tfrac{1}{4})\varepsilon.$$

Similarly, $t_3 = t_2 \wedge s_3 = t_2 + s_3 - t_2 \vee s_3$, so that

$$\int_a^b t_3(x)\, dx > \left(\int_a^b h_2(x)\, dx - (\tfrac{1}{2} + \tfrac{1}{4})\varepsilon \right) + \int_a^b h_3(x)\, dx - \tfrac{1}{8}\varepsilon - \int_a^b h_2(x)\, dx$$

$$= \int_a^b h_3(x)\, dx - (\tfrac{1}{2} + \tfrac{1}{4} + \tfrac{1}{8})\varepsilon.$$

An obvious induction now shows that for all n,

$$\int_a^b t_n(x)\, dx > \int_a^b h_n(x)\, dx - \left(\tfrac{1}{2} + \tfrac{1}{4} + \cdots + \frac{1}{2^n} \right)\varepsilon$$

$$> \int_a^b h_n(x)\, dx - \varepsilon.$$

Hence $\int_a^b h_N(x)\,dx < \int_a^b t_N(x)\,dx + \varepsilon < 2\varepsilon$, and the result follows.

We finish this section with a result which shows when a sequence of functions can be differentiated termwise. Example A.22(ii) shows that uniform convergence of a sequence (f_n) of differentiable functions does not guarantee a differentiable limit, while the trivial example $f_n(x) = n$ for all real x shows that (f'_n) may converge while (f_n) does not. These examples lead us to the following result.

Theorem A.31 Let (f_n) be a sequence of real- or complex-valued functions on a bounded open interval $I = (a, b)$. Suppose that the sequence (f'_n) is uniformly convergent on I, and that for some x_0 in I, the sequence $(f_n(x_0))$ is convergent. Then the sequence (f_n) is uniformly convergent on I, and its limit function f, say, is differentiable and $f' = \lim_{n \to \infty} (f'_n)$.

Proof We can suppose the functions involved are real-valued without loss of generality. For a given point t in I, define a sequence (g_n) by

$$g_n(x) = \begin{cases} [f_n(x) - f_n(t)]/(x - t) & \text{if } x \neq t, \\ f'_n(t) & \text{if } x = t. \end{cases}$$

Then for any m, n,

$$g_n(x) - g_m(x) = \frac{[f_n(x) - f_n(x)] - [f_n(t) - f_m(t)]}{x - t}$$

$$= f'_n(x_n) - f'_m(x_1)$$

for some x_1 between t and x, by the mean-value theorem, applied to $(f_n - f_m)$. The general principle of convergence (see remark following Definition A.21) now shows that the sequence (g_n) is uniformly convergent on I. In particular, taking $t = x_0$, we have $f_n(x) = f_n(x_0) + (x - x_0)g_n(x)$, and since $(f_n(x_0))$ is convergent, and (g_n) is uniformly convergent, it follows that (f_n) is uniformly convergent (the boundedness of I is used at this point – how?) Let f denote the limit of the sequence (f_n) whose existence we have just established. Returning to a general point t, the definition of g_n shows that f is continuous at t, and hence that if $g_n \to g$ as $n \to \infty$, g is also continuous at t (Theorem A.25(i)), i,e, $g(x) \to g(t)$ as $x \to t$.
But for $x \neq t$, $g(x) = \lim_{n \to \infty} g_n(x) = [f(x) - f(t)]/(x - t)$, while if $x = t$, $g(t) = \lim_{n \to \infty} g_n(t) = \lim_{n \to \infty} f'_n(t)$. Hence the statement that $g(x) \to g(t)$ as $x \to t$ shows that f is differentiable at t, and that $f'(t)$ exists and is equal to $\lim_{n \to \infty} f'_n(t)$, as required.

Our next result is a useful fact about differentiation under the integral sign.

176

176

Theorem A.32 Let f be a continuous real- or complex-valued function on the rectangle $R = (a, b) \times [c, d]$. For $a < x < b$ define

$$F(x) = \int_c^d f(x, y) \, dy.$$

If f has a partial derivative f_1 with respect to x which exists and is continuous at all points of R, then F is differentiable on (a, b) and

$$F'(x) = \int_c^d f_1(x, y) \, dy \text{ for } a < x < b.$$

Proof Let us fix a point x_1 in (a, b) and then choose a_1, b_1 with $a < a_1 < x_1 < b_1 < b$. Then f_1 will be continuous on the closed bounded rectangle $R_1 = [a_1, b_1] \times [c, d]$, and so will be uniformly continuous by Lemma A.4 (extended to functions on R_1: the proof uses the result on sub-sequences in the same way). It follows that for a given $\varepsilon > 0$, there will be some $\delta > 0$ such that $|f_1(x, y) - f_1(x', y')| < \varepsilon$ if (x, y), (x', y') are in R_1 and are separated by at most δ.

If we choose h so that $x_1 + h$ is in $[a_1, b_1]$, we can write

$$F(x_1 + h) - F(x_1) = \int_c^d (f(x_1 + h, y) - f(x_1, y)) \, dy$$

$$= h \int_c^d f_1(x_1 + \theta h, y) \, dy, \qquad \text{for some } \theta \text{ in } (0, 1)$$

(depending on x, h, and y) by the mean-value theorem.

It follows that if $|h| < \delta$, we have

$$\left| \frac{F(x_1 + h) - F(x_1)}{h} - \int_c^d f_1(x_1, y) \, dy \right| = \left| \int_c^d (f_1(x_1 + \theta h, y) - f_1(x, y)) \, dy \right|$$

$$\leqslant \int_c^d |f_1(x_1 + \theta h, y) - f_1(x_1, y)| \, dy$$

$$\leqslant \int_c^d \varepsilon \, dy = \varepsilon(d - c).$$

Hence F is differentiable at x_1, and $F'(x_1) = \int_c^d f_1(x_1, y) \, dy$, and the result follows since x_1 was any point of (a, b).

A.4 DOUBLE SERIES AND INTEGRALS

In this section we outline briefly the results which we shall need concerning double series and integrals. Since we need only rather restricted special cases, we shall not attempt any great generality. We begin with the integral of a continuous function over a rectangle.

Theorem A.33 Let f be continuous on the rectangle $R = [a, b] \times [c, d]$, and define functions f_1, f_2 on $[c, d], [a, b]$ respectively by

$$f_1(y) = \int_a^b f(x, y)\, dx, \qquad f_2(x) = \int_c^d f(x, y)\, dy.$$

Then both f_1 and f_2 are continuous on their domains, and

$$\int_c^d f_1(y)\, dy = \int_a^b f_2(x)\, dx.$$

Proof The continuity of f_1, for example, is immediate from

$$|f_1(y) - f_1(y')| \leqslant \int_a^b |f(x, y) - f(x, y')|\, dx,$$

and it remains to prove the equality of the integrals of f_1 and f_2.

To do this we define new functions F and G on R by

$$F(x, y) = \int_a^x \left(\int_c^y f(u, v)\, dv \right) du, \qquad G(x, y) = \int_c^y \left(\int_a^x f(u, v)\, du \right) dv,$$

where the existence of the integrals is shown by the preceding argument. The required result is that $F(b, d) = G(b, d)$. On differentiation we obtain $\partial F/\partial x = \int_c^y f(x, v)\, dv$ by the fundamental theorem of calculus, while $\partial G/\partial x = \int_c^y f(x, v)\, dv = \partial F/\partial x$ by Theorem A.32. It follows that $F - G$ is constant on any line parallel to the x-axis, and a similar argument shows that $F - G$ is constant on any line parallel to the y-axis, and is therefore constant on R. Hence

$$F(b, d) - G(b, d) = F(a, c) - G(a, c) = 0, \text{ as required.}$$

Corollary A.34 The above result remains true

(a) when f is required only to be bounded and continuous on the open rectangle $(a, b) \times (c, d)$, and
(b) when f is bounded on $[a, b] \times [c, d]$ and continuous except at points which lie on a finite number of lines parallel to the axes.

Proof To prove (a) it is only necessary to remove a small strip around the perimeter of the rectangle, on which, since f is bounded, the relevant integrals will be small, and apply Theorem A.33 to the rest of the rectangle. To prove (b), divide the rectangle into a finite number of subrectangles, to which (a) applies.

Definition A.35 The common value of the integrals in Theorem A.33 (subject to the hypotheses of Theorem A.33 or Corollary A.34) is called the *integral of f over* R, and will be denoted

$$\int_R f(x, y)\, d(x, y)$$

in addition to the expressions already derived in terms of repeated integrals. (A more complete treatment would also include an 'intrinsic' definition of the double integral by means of dissections and a limiting process analogous to Definition A.9. For this the reader can refer to the books on analysis in the Bibliography.

In Chapter 5, we required another extension of the double integral to the case when the integrals have infinite range, in situations when the integrand $f(x, y)$ satisfies a bound of the form $\int_a^b \int_c^d |f(x, y)| \, \mathrm{d}(x, y) \leqslant M$, where M is independent of a, b, c, d.

We leave it to the reader to supply the simple modification of the above definitions to accommodate this possibility.

Our second topic in this section is that of double sequences and series.

Definition A.36 We say S is a (real- or complex-valued) *double sequence* if for each $m, n = 0, 1, 2, \ldots, S(m, n)$ is a real or complex number.

We say that the double sequence converges to l as $m, n \to \infty$ if for each $\varepsilon > 0$, we can find N such that $|S(m, n) - l| < \varepsilon$, when *both* m and n are $\geqslant N$.

In this case we write $\lim_{m, n \to \infty} S(m, n) = l$, and call l the *double limit* of S.

We must immediately distinguish the double limit as just defined, from the *iterated limits* defined by $\lim_{m \to \infty} \left(\lim_{n \to \infty} S(m, n) \right)$ or $\lim_{n \to \infty} \left(\lim_{m \to \infty} S(m, n) \right)$. The difference is illustrated by the following examples.

Example A.37 (i) Let $S(m, n) = mn/(m^2 + n^2)$ for $m, n \geqslant 1$. Then the iterated limit $\lim_{n \to \infty} (\lim_{m \to \infty} S(m, n))$ exists and equals zero, while the double limit does not exist; for instance $S(n, n) = \frac{1}{2}$, $S(n, 2n) = 2/5$ for all n.

(ii) Let $S(m, n) = (-1)^{m+n} \left(\dfrac{1}{m} + \dfrac{1}{n} \right)$.

Then the double limit exists and is zero, since for $\varepsilon > 0$ we need only take both m, $n > 2\varepsilon^{-1}$, while the limits $\lim_{n \to \infty} S(m, n)$ or $\lim_{m \to \infty} S(m, n)$ do not exist, so the iterated limits cannot exist either.

We leave it as an easy exercise to show that if the double limit and one (or both) of the iterated limits exist, then they must have the same value. One can also show that a double sequence which is monotone in both m, n separately is convergent if and only if it is bounded.

Definition A.38 Let $t(m, n)$ be a double sequence. The double sequence defined by $S(m, n) = \sum_{p=1}^{m} \sum_{q=1}^{n} t(p, q)$ is called a *double series*. We shall say that the double series is convergent with sum S if and only if the double limit

$$\lim_{m,n\to\infty} S(m,n) \text{ exists and is equal to } S, \text{ and we write}$$

$$\sum_{m,n=1}^{\infty} t(m,n) = S \text{ in this case.}$$

This apparently natural definition has some surprising consequences. Firstly, it goes without saying that the convergence of the double series must be distinguished from the iterated sums

$$\sum_{n=1}^{\infty}\left(\sum_{m=1}^{\infty} t(m,n)\right) \quad \text{and} \quad \sum_{m=1}^{\infty}\left(\sum_{n=1}^{\infty} t(m,n)\right),$$

as Examples A.37 show. Secondly, while it is certainly the case that the convergence of the double sum implies that $t(m,n)\to 0$ as $m,n\to\infty$ (the reader should prove this while remembering that the converse is false), it need not imply that, for instance, $\lim_{n\to\infty} t(m,n)$ exists and is zero for all m. For if $t(m,n)=1$ if $\max(m,n)=1$, $t(m,n)=-1$ if $\max(m,n)=2$, $t(m,n)=0$ otherwise, then the double sum converges (its sum is 2) though the indicated limits are not all zero. However, we shall have little use for these oddities, since in Chapter 6 our series are absolutely convergent.

Definition A.39 Let $S(m,n)=\sum_{p=1}^{m}\sum_{q=1}^{n} t(p,q)$ be a double series as in Definition A.38.

We shall say that the series is *absolutely convergent* if the series $\sum\sum|t(p,q)|$ is convergent. By the remark preceding Definition A.38, it is equivalent to assume the boundedness of the partial sums $\sum_{p=1}^{m}\sum_{q=1}^{n}|t(p,q)|$.

The most importance property of absolutely convergent series is as follows.

Theorem A.40 Let $S(m,n)=\sum_p\sum_q t(p,q)$ be an absolutely convergent series.
Then (i) $S(m,n)$ is convergent, and
(ii) the iterated sums exist, and are equal to the double sum.

To prove this result it is interesting to show that absolute convergence of a series is independent of the order structure of the set of integers on which the series is defined (this remark applies equally to ordinary and multiple series). We begin by exploring this situation, and return to the proof of Theorem A.40 later.

Definition A.41 Let S be any set, and f a real- or complex-valued function on S. We say that the sum $\sum_{s\in S} f(s)$ is (unordered) *convergent* if there is a number A such that for each $\varepsilon>0$, there is a finite subset F of S, such that for any finite subset F' of S with $F'\supseteq F$, we have $|\sum_{s\in F'} f(s)|\sum_{s\in F'} f(s) - A|<\varepsilon$. We say that the sum $\sum_{s\in S} f(s)$ is *absolutely convergent* if there is an upper bound M, say, for all sums of the form $\sum_{s\in F}|f(s)|$.

We now show that these concepts are in fact equivalent.

Lemma A.42 The sum $\sum_{s\in S} f(s)$ is unordered convergent if and only if it is absolutely convergent. In this case the set $\{s: f(s) \neq 0\}$ must be denumerable (but this is plainly not sufficient for convergence!)

Proof Write $f = g + ih$ where g, h are real valued, and $g = g_1 - g_2$, $h = h_1 - h_2$, where $g_1 = \max(g, 0)$, $g_2 = \max(-g, 0)$, and similarly for h.

If we first suppose that $\sum f(s)$ is absolutely convergent then since, for example, $0 \leqslant g_1 \leqslant |g| \leqslant |f|$, it follows that $\sum g_1(s)$, etc., are also absolutely convergent. Let $G_1 = \sup\{\sum_{s\in F} g_1(s): F$ a finite subset of $S\}$, and define G_2, H_1, H_2 similarly.

We shall show that $\sum f(s)$ is unordered convergent with sum $G_1 - G_2 + i(H_1 - H_2)$: at this point the reader should pause to see why the definition is linear in f, so that it is sufficient to show that $\sum g_1(s)$ is unordered convergent with sum G_1.

But given $\varepsilon > 0$, there is a subset F_1 of S such that

$$G_1 - \varepsilon < \sum_{s\in F_1} g_1(s) \leqslant G_1,$$

and since g_1 is positive this holds for any finite subset $F' \supseteq F_1$, as required.

Suppose conversely that $\sum_{s\in S} f(s)$ is unordered convergent, with sum $A = A_1 + iA_2$.

Then for any finite set F,

$$\left| \sum_{s\in F} g(s) - A_1 \right| = \left| \mathrm{Re}\left(\sum_{s\in F} f(s) - A \right) \right| \leqslant \left| \sum_{s\in F} f(s) - A \right|,$$

and it follows that $\sum g(s)$ (and $\sum h(s)$ similarly) are unordered convergent with sum A_1 (or A_2). We complete the argument by showing that $\sum g(s)$ must be absolutely convergent.

Let S_1, S_2 denote the sets on which $g(s) > 0$, $g(s) < 0$ respectively, so that $S_1 \cap S_2 = \phi$ and $g = g_1$ on S_1, $g = -g_2$ on S_2.

Then taking $\varepsilon = 1$ in the definition, there is a fixed finite set F in S such that

$$A_1 - 1 < \sum_{s\in F'} (g_1(s) - g_2(s)) < A_1 + 1$$

for any finite $F' \supseteq F$.

Write $F_1 = F \cap S_1$, $F_2 = F \cap S_2$.

Then for any finite subset K of S_1,

$$\sum_{s\in K} g_1(s) \leqslant \sum_{s\in K\cup F_1} g_1(s) = \sum_{s\in K\cup F} g_1(s)$$
$$< A_1 + 1 + \sum_{s\in K\cup F} g_2(s) = A_1 + 1 + \sum_{s\in F_2} g_2(s),$$

and the fixed quantity on the right gives an upper bound for $\sum_{s\in K} g_1(s)$.

Consequently $\sum g_1(s)$ is absolutely convergent: a similar argument gives the same result for $\sum g_2(s)$. The proof of Lemma A.42 is completed by the observation that if $\sum_{s\in F} |g(s)| < M$ for any finite set F, then for any $n = 1, 2, \ldots$, the number of points where $|g(s)| > 1/n$ is at most nM, and thus the set where $g(s) \neq 0$ is a denumerable union of finite sets, and is thus itself denumerable.

Proof of Theorem A.40 We suppose that $\sum\sum t(p, q)$ is absolutely convergent: it follows that all the single series such as $\sum_{p=1}^{\infty} t(p, q)$ are again absolutely convergent. The set S on which the double series is defined is here $\mathbb{N} \times \mathbb{N}$, the product of two copies of the natural numbers, and any finite subset of this can be contained in a 'rectangle' of the form $\{1, 2, 3, \ldots, m\} \times \{1, 2, 3, \ldots, n\}$. By taking sets of this form in the definition of unordered convergence (and applying Definition A.42) it follows that the double series is convergent – a similar observation shows all the single series are convergent, too.

More precisely, denoting the sum of the series by T, we have that, for given $\varepsilon > 0$, we can find N such that

$$|S(m, n) - T| = \left| \sum_{p=1}^{m} \sum_{q=1}^{n} t(p, q) - T \right| < \varepsilon \text{ if } m, n \text{ are both } \geqslant N.$$

But since the single series are convergent we can let m (or n) $\to \infty$ while keeping the other fixed, so that we obtain, for example

$$\left| \sum_{p=1}^{m} \sum_{q=1}^{\infty} t(p, q) - T \right| \leqslant \varepsilon \text{ if } m \geqslant N,$$

and the iterated sums have the same value as the double series.

When we come to apply these results in Chapter 6, we principally require the comparison test to deduce the absolute convergence of a given series from a known standard series. For single series, a standard of comparison is often the series $\sum_{n=1}^{\infty} 1/n^{\alpha}$ for $\alpha > 1$. The corresponding result for double series is as follows.

Lemma A.43 The series $\sum\sum_{m, n=1}^{\infty} (m^2 + n^2)^{-\alpha}$ is convergent if $\alpha > 1$, divergent if $\alpha \leqslant 1$.

Proof We establish the convergence or divergence by comparison with corresponding integrals – a common technique in the theory of ordinary series. Suppose firstly that $\alpha > 1$. The function $f(x, y) = (x^2 + y^2)^{-\alpha}$ is decreasing in x and y, so that for each m, n, we have

$$f(m, n) \leqslant \int_{m-1}^{m} \int_{n-1}^{n} f(x, y) \, \mathrm{d}(x, y)$$

where the notation is as in Definition A.35.

If we do this for each term except for $m = n = 1$, we obtain

$$\sum_{m=1}^{N} \sum_{n=1}^{N} f(m, n) \leqslant f(1, 1) + \int_{T} f(x, y) \, \mathrm{d}(x, y)$$

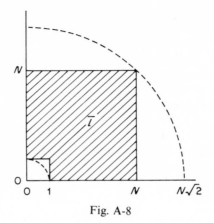

Fig. A-8

where T is the set $[0, N] \times [0, N] \setminus [0, 1] \times [0, 1]$ indicated in Fig. A.8. To estimate the integral over T we further increase the region up to the circular arcs of radius 1 and $\sqrt{2N}$ as indicated, and change to polar co-ordinates.

This gives

$$\sum_{m=1}^{N} \sum_{n=1}^{N} (m^2 + n^2)^{-\alpha} \leqslant \frac{1}{2^{\alpha}} + \int_0^{\pi/2} \left(\int_1^{\sqrt{2N}} (r^2)^{-\alpha} r \, dr \right) d\theta$$

$$= \frac{1}{2^{\alpha}} + \frac{\pi}{2} \left[\frac{r^{2-2\alpha}}{2-2\alpha} \right]_1^{\sqrt{2N}} < \frac{1}{2^{\alpha}} + \frac{\pi}{4(\alpha - 1)},$$

a bound which is independent of N. It follows that the series is convergent, as required.

To show divergence when $\alpha \leqslant 1$ it is obviously sufficient to consider $\alpha = 1$ alone. This time

$$f(m, n) = (m^2 + n^2)^{-1} \geqslant \int_m^{m+1} \int_n^{n+1} f(x, y) \, d(x, y)$$

so that

$$\sum_{m=1}^{N} \sum_{n=1}^{N} f(m, n) \geqslant \int_1^{N+1} \int_1^{N+1} f(x, y) \, d(x, y).$$

It is possible to change to polar co-ordinates as before, but we follow a different route for the sake of variety.

On integrating directly, we obtain

$$\int_1^{N+1} \int_1^{N+1} \frac{dx \, dy}{x^2 + y^2} = \int_1^{N+1} \left[\frac{1}{y} \tan^{-1} \frac{x}{y} \right]_1^{N+1} dy$$

$$= \int_1^{N+1} \frac{1}{y} \tan^{-1} \left\{ \frac{(N+1)/y - 1/y}{1 + (N+1)/y^2} \right\} dy$$

$$= \int_1^{N+1} \frac{1}{y} \tan^{-1} \left\{ \frac{Ny}{y^2 + N + 1} \right\} dy.$$

On the range $[1, N+1]$ the function $Ny/(y^2 + N + 1)$ has a maximum when $y^2 = N + 1$ and a minimum of $N/(N+2)$ at the end points.

Hence the above integral is strictly greater than

$$\tan^{-1}\left(\frac{N}{N+2}\right)\int_1^{N+1}\frac{dy}{y}$$

which tends to infinity with N as required.

APPENDIX B

Some Definite Integrals

We collect in this appendix a number of useful special integrals which are needed in the study of Fourier series and integrals.

Lemma B.1 Let $I_n = \int_0^{\pi/2} (\cos x)^n \, dx$ for $n = 0, 1, 2, \ldots$, Then

$$I_n = \int_0^{\pi/2} (\sin x)^n \, dx = \begin{cases} \dfrac{(n-1)(n-3)\ldots 3 \cdot 1}{n(n-2)\ldots 4 \cdot 2} \dfrac{\pi}{2} & \text{if } n \text{ is even,} \\[2ex] \dfrac{(n-1)(n-3)\ldots 4 \cdot 2}{n(n-2)\ldots 5 \cdot 3} \end{cases}$$

Proof The fact that $I_n = \int_0^{\pi/2} (\cos x)^n \, dx = \int_0^{\pi/2} (\sin x)^n \, dx$ is immediate on putting $\pi/2 - x$ for x. One finds on integrating by parts that if $n \geqslant 2$,

$$I_n = \int_0^{\pi/2} (\cos x)^{n-1} (\cos x) \, dx = [(\cos x)^{n-1} \sin x]_0^{\pi/2}$$

$$+ \int_0^{\pi/2} (n-1)(\cos x)^{n-2} (\sin x)^2 \, dx$$

$$= 0 + (n-1) \int_0^{\pi/2} (\cos x)^{n-2} (1 - \cos^2 x) \, dx$$

$$= (n-1)(I_{n-2} - I_n).$$

On rearranging we obtain the reduction formula

$$I_n = \frac{n-1}{n} I_{n-2},$$

and the result is immediate from this and the special values $I_0 = \pi/2$, $I_1 = 1$.

Lemma B.2 Let I_n be as in Lemma B.1. Then $\sqrt{n} I_n \to \sqrt{\tfrac{1}{2}\pi}$ as $n \to \infty$.

184

Proof Since $0 \leqslant \cos x \leqslant 1$ for x in $[0, \pi/2]$, we have $I_{n<1} < I_n < I_{n-1}$ for all n.
Suppose that n is odd. Then

$$I_{n+1} = \frac{n}{n+1} \cdot \frac{n-2}{n-1} \cdot \dots \cdot \frac{3}{4} \cdot \frac{1}{2} \cdot \frac{\pi}{2}$$

$$< I_n = \frac{n-1}{n} \cdot \frac{n-3}{n-2} \cdot \dots \cdot \frac{4}{5} \cdot \frac{2}{3}$$

$$< \frac{n-2}{n-1} \cdot \dots \cdot \frac{3}{4} \cdot \frac{1}{2} \cdot \frac{\pi}{2} = I_{n-1}.$$

Hence

$$\frac{1}{n+1} \cdot \frac{\pi}{2} < I_n^2 < \frac{1}{n} \cdot \frac{\pi}{2},$$

and the result follows in this case.

A similar argument holds if n is even.

Lemma B.3 Let

$$f_n(x) = \begin{cases} (1 - x/n)^n & \text{if } 0 \leqslant x \leqslant n \\ 0 & \text{if } x > n \end{cases} \quad \text{for } n = 1, 2, 3, \dots.$$

Then $f_n(x)$ is monotone increasing, with limit e^{-x} as $n \to \infty$.

Proof The fact that the limit is e^{-x} is a well-known consequence of the binomial expansion of $(1 - x/n)^n$. To show monotonicity consider $g(t) = (1 - x/t)^t$ for $t > x$. We have

$$\frac{g'(t)}{g(t)} = \log\left(1 - \frac{x}{t}\right) + t\left(1 - \frac{x}{t}\right)^{-1} \frac{x}{t^2} = \log \theta + (1 - \theta)/\theta,$$

where $\theta = 1 - x/t$ is between 0 and 1.
But for $y > 1$,

$$\log y = \int_1^y \frac{du}{u} < y - 1,$$

so on putting $y = 1/\theta$, we have $-\log \theta < 1/\theta - 1$, or $\log \theta + (1 - \theta)/\theta > 0$ for $0 < \theta < 1$.
Hence $g'(t) > 0$, and g is increasing, as required.

Theorem B.4 $\int_{-\infty}^{\infty} e^{-x^2} \, dx = \sqrt{\pi}$, or equivalently $\int_{-\infty}^{\infty} e^{-\pi t^2} \, dt = 1$.

Proof The integral is convergent since, for instance,

$$\int_1^a e^{-x^2} \, dx < \int_1^a e^{-x} \, dx = e^{-1} - e^{-a}.$$

186

Hence given $\varepsilon > 0$, we may choose X such that

$$\int_0^\infty e^{-x^2}\,dx > \int_0^X e^{-x^2}\,dx > \int_0^\infty e^{-x^2}\,dx - \varepsilon.$$

But on $[0, X]$, $(1 - x^2/n)^n$ increases to e^{-x^2}, as $n \to \infty$, and by Dini's theorem (Theorem A.26) the convergence must be uniform. Hence for sufficiently large n we have

$$\int_0^\infty e^{-x^2}\,dx > \int_0^{\sqrt{n}}(1 - x^2/n)^n\,dx > \int_0^X (1 - x^2/n)^n\,dx > \int_0^\infty e^{-x^2}\,dx - 2\varepsilon.$$

But $\int_0^{\sqrt{n}}(1 - x^2/n)^n\,dx = \sqrt{n}\int_0^{\pi/2}(\cos\theta)^{2n+1}\,d\theta,$ putting $x = \sqrt{n}\sin^2\theta,$

$$= \sqrt{n}\,I_{2n+1}$$

$$= \sqrt{\frac{n}{2n+1}} \cdot \sqrt{2n+1}\,I_{2n+1}$$

$$\to \frac{1}{\sqrt{2}} \cdot \sqrt{\frac{\pi}{2}} = \tfrac{1}{2}\sqrt{\pi} \qquad \text{from Lemma B.3.}$$

Hence $\int_0^\infty e^{-x^2}\,dx = \tfrac{1}{2}\sqrt{\pi}$, and the result is proved.

Theorem B.5 (i) $\displaystyle\int_{-\infty}^\infty \frac{\sin^2\theta}{\theta^2} = \pi,$

(ii) $\displaystyle\lim_{a\to\infty}\int_0^a \frac{\sin\theta}{\theta}\,d\theta = \tfrac{1}{2}\pi.$

Proof (i) We know that $\sin^2\theta$ is periodic with period π, so that

$$\int_{-\infty}^\infty \frac{\sin^2\theta}{\theta^2}\,d\theta = \sum_{n=-\infty}^\infty \int_{n\pi}^{(n+1)\pi} \frac{\sin^2\theta}{\theta^2}\,d\theta$$

$$= \sum_{n=-\infty}^\infty \int_0^\pi \frac{\sin^2\theta}{(\theta - n\pi)^2}\,d\theta,$$

putting $\theta - n\pi$ for θ.

From exercise 4 of chapter 1 (a corollary of Example 1.12(ii)) we know that

$$\frac{\pi^2}{\sin^2 \pi x} = \sum_{-\infty}^\infty \frac{1}{(x-n)^2},$$

or

$$\sum_{-\infty}^\infty \frac{\sin^2\theta}{(\theta - n\pi)^2} = 1, \qquad \text{putting } \pi x = \theta.$$

But each term in this series is positive and continuous on $[0, \pi]$, so that by Dini's theorem (Theorem A.26) the convergence is uniform on $[0, \pi]$. Hence by Theorem A.27 we may integrate termwise to obtain

$$\sum_{-\infty}^{\infty} \int_0^{\pi} \frac{\sin^2 \theta}{(\theta - n\pi)^2} \, d\theta = \int_0^{\pi} \sum_{-\infty}^{\infty} \frac{\sin^2 \theta}{(\theta - n\pi)^2} \, d\theta$$

$$= \int_0^{\pi} d\theta = \pi, \text{ as required.}$$

(ii) We integrate by parts to obtain

$$\int_0^a \frac{\sin \theta}{\theta} \, d\theta = \left[\frac{1 - \cos \theta}{\theta} \right]_0^a + \int_0^a \left(\frac{1 - \cos \theta}{\theta^2} \right) d\theta$$

$$= \frac{1 - \cos a}{a} + \int_0^a \frac{2 \sin^2 \frac{1}{2}\theta}{\theta^2} \, d\theta,$$

since $(1 - \cos \theta)/\theta \to 0$ as $\theta \to 0$.

Hence letting $a \to \infty$, we find that

$$\lim_{a \to \infty} \int_0^a \frac{\sin \theta}{\theta} \, d\theta$$

exists and is equal to

$$\int_0^{\infty} 2 \frac{\sin^2 \frac{1}{2}\theta}{\theta^2} \, d\theta = \int_0^{\infty} \frac{\sin^2 \theta}{\theta^2} \, d\theta = \frac{\pi}{2} \text{ from (i).}$$

Notice that in (ii) the integral does not satisfy condition (ii) of Definition A.13 – in fact

$$\int_{n\pi}^{(n+1)\pi} \frac{|\sin \theta|}{\theta} \, d\theta > \frac{1}{(n+1)\pi} \int_{n\pi}^{(n+1)\pi} |\sin \theta| \, d\theta = \frac{2}{(n+1)\pi}$$

so that $\int_a^b |\sin \theta|/\theta \, d\theta$ is unbounded. Hence we write

'$\lim_{a \to \infty} \int_0^a$' rather than the more natural '$\int_0^{\infty}$'.

Theorem B.6 For $0 < a < 1$, $\int_0^{\infty} t^{a-1}/(1 + t) \, dt = \pi/\sin a\pi$.

Proof The conventional method of evaluation of this integral is to use the theory of residues. We outline an alternative method more in line with the results above, leaving the reader to fill in the details.

Consider first the integral over $[0, 1]$. For $0 \leqslant t \leqslant 1$, put $s_n(t) = 1 - t + t^2 + \cdots + (-t)^n$, and notice that $0 \leqslant s_n(t) \leqslant 1$ for all $n = 0, 1, 2, 3, \ldots$. Evidently

$$\int_0^1 t^{a-1} s_n(t) \, dt = \sum_{k=0}^n \frac{(-1)^k}{a + k}$$

188

on integrating termwise. For any $\delta < 1$, $s_n(t) \to (1 + t)^{-1}$ uniformly on $[0, \delta]$, while the uniform boundedness of $s_n(t)$ shows that both

$$\int_\delta^1 t^{k-1} s_n(t)\,dt \quad \text{and} \quad \int_\delta^1 \frac{t^{a-1}}{1+t}\,dt$$

may be made as small as required by taking δ near to 1. It follows that we can let $n \to \infty$ to obtain

$$\int_0^1 \frac{t^{a-1}}{1+t}\,dt = \sum_0^\infty \frac{(-1)^k}{a+k}.$$

A similar argument shows that

$$\int_1^\infty \frac{t^{a-1}}{1+t}\,dt = \sum_0^\infty \frac{(-1)^{k+1}}{a-k-1},$$

and it follows that

$$\int_0^\infty \frac{t^{a-1}}{1+t}\,dt = \sum_{-\infty}^\infty \frac{(-1)^k}{a+k} = \frac{\pi}{\sin a\pi},$$

using the result of Example 1.12(ii).

Appendix C

Bibliography and Guide to Further Reading

We have assumed that the reader has already had a first course in mathematical analysis. The following two books contain more than sufficient by way of prerequisites for the present book:

M. Spivak (1967) *Calculus*, Benjamin;

K. A. Ross (1980) *Elementary Analysis*, Springer.

For the more advanced theory including real and complex analysis and the Lebesgue integral, one could begin with:

R. G. Bartle (1976) *The Elements of Real Analysis*, J. Wiley;

W. Rudin (1964) *Principles of Mathematical Analysis* (2nd edn), McGraw-Hill;

G. J. O. Jameson (1974) *Topology and Normed spaces*, Chapman & Hall;

L. V. Ahlfors (1966) *Complex Analysis*, McGraw-Hill.

The classic account of summability processes is in:

G. H. Hardy (1963) *Divergent Series*, Oxford.

Fourier theory itself is treated at a fairly elementary level in the following (which do not use the Lebesgue integral):

W. Rogosinski (1950) *Fourier Series* (trans. Cohen and Steinhardt), Chelsea (NY);

C. Lanczos (1966) *A Discourse on Fourier Theory*. Oliver and Boyd. London.

W. E. Byerly (1959) *An Elementary Treatise on Fourier's Series* Dover.

For the classical viewpoint on trigonometric series one must begin with

A. Zygmund (1959) *Trigonometric Series*, CUP, or with

Y. Katznelson (1968) *An Introduction to Harmonic Analysis*, J. Wiley, which puts the theory into the wider context of functional analysis and topological groups.

For Fourier integrals, we mention

E. M. Stein and G. Weiss (1971) *An Introduction to Fourier Analysis on Euclidean Spaces*, Princeton,

which contains in addition an attractive presentation of the theory of distributions.

A highly stimulating account which makes considerable demands on the reader's analytical technique, is contained in

  H. Dym and H. P. McKean (1972) *Fourier Series and Integrals*, Academic Press,

which is unique in its approach to both theory and practice. For an ambitious reader it would make the ideal sequel to the present book.

There are substantial works on mathematical physics, particularly partial differential equations, which use Fourier theory extensively. We mention, among many:

  H. Bateman (1932) *Partial Differential Equations of Mathematical Physics*, CUP,

  H. Jeffreys and B. S. Jeffreys (1956) *Methods of Mathematical Physics* (3rd edn), CUP.

The books which apply Fourier theory to investigate specific areas in physics or engineering are too numerous for any coherent account to be given. The author's first encounter was with the solution of the wave equation in acoustical theory in

  H. Lamb (1960) *The Dynamical Theory of Sound*, Dover.

Applications in vibration theory are treated in

  D. E. Newland (1975) *An Introduction to Random Vibrations and Spectral Analysis*, Longman.

and the problems of static stresses in plates in

  R. Szilard (1973) *Theory and Analysis of Plates*, Prentice Hall.

An extended list of special transforms can be found in

  A. Erdelyi *et al.* (1954) *Tables of Integral Transforms*, McGraw-Hill.

Other books which the author found interesting include:

  E. Hecht and A. Zajac (1974) *Optics*, Addision Wesley;

  R. N. Bracewell (1978) *The Fourier Transform and its Applications* (2nd edn), McGraw-Hill;

but these are little more than random selections from an enormous variety of uses to which the theory has been put. Indeed, at this point it is the reader's own needs and preferences which should guide his or her further study.

Index

WITHDRAWN